HOW TO RESTORE
TRACTOR
MAGNETOS

Neil Yerigan

MBI Publishing Company

Acknowledgements

I dedicate this book to Joseph Malin of Standard Magneto Sales Co., Inc., Chicago, Illinois, and to Sally Burns of Bloomington, Minnesota. Without the advice and information freely given by Joe Malin over the last 20 years, I would have quit the magneto repair business long ago. Without the encouragement of Sally Burns, I would have never even started writing this book.

Black and white photographs were processed and printed by Mark Jensen, Minneapolis, Minnesota

First published in 1994 by MBI Publishing Company, PO Box 1, 729 Prospect Avenue, Osceola, WI 54020-0001 USA

© Neil C. Yerigan, 1994

MBI Publishing Company books are also available at discounts in bulk quantity for industrial or sales-promotional use. For details write to Special Sales Manager at Motorbooks International Wholesalers & Distributors, 729 Prospect Avenue, PO Box 1, Osceola, WI 54020-0001 USA

Library of Congress Cataloging-in-Publication Data

Yerigan, Neil C.
    How to restore tractor magnetos/Neil C. Yerigan.
        p.   cm.
    Includes index.
    ISBN 0-87938-949-4
    1. Magneto—Maintenance and repair—Amateurs' manuals. 2. Farm tractors—Conservation and restoration—Amateurs' manuals. I. Title. II. Title: How to restore tractor magnetos.
TL213.Y47   1994
629.25'04–dc20   9432972
                                      CIP

On the front cover: One of the best-known tractors of all time—the John Deere B. *Andrew Morland*

On the back cover: Top: a rebuilt International-Havester F4 magneto on a Farmall F-12 tractor. Bottom: another International F4 magneto with the cover removed.

Printed in the United States of America

Contents

Introduction

One of my good customers, some years ago, affixed me with a beady stare and said, "Who is going to replace you, when you are gone?"

"Gone?" I replied. "I'm not going anywhere."

He gazed pointedly at my somewhat portly tummy. "We had better face facts. Neither of us is getting any younger. Or healthier."

I thought it over for a minute or so. "What choice do I have? I'm not going to hire anyone, like an apprentice. Too many problems, like quarterly reports, workers compensation, unemployment compensation, OSHA." My nose began to wrinkle up, as if there was a bad smell in the air. But the air smelled the same as always; a friendly mixture of grease and solvent. "Maybe I should write a book," I said.

Several years have passed and here is the book. In it, I do my best to explain how a magneto works and how you can repair one. I begin by reviewing basic electricity and explaining how magnetism interacts with conductors, and how electrical current produces a magnetic field. I also explain the difference between a capacitor and an inductor and how they affect an electric current. If you understand these things, you will be able to understand how a magneto produces a spark.

I try to tell you everything about magnetos that you ever wanted to know, and then some. Af-

ter that, I give you a distillation of thirty years experience in fixing old junk in the chapter on "Non-destructive Dismantling of Antique Magnetos." I had this chapter looked over by my old friend Frank Bethard, who spent a lifetime restoring old cars and airplanes. He made some comments, returned my chapter, and suggested adding, "Patience is the most important tool." I agree.

I also tell you more about meters and gauges than you really want to know. I used to grumble about excessive detail on meters, but, in the end, I appreciated the fact that learning how to measure volts, ohms, and amps teaches you a lot about electricity.

I thoroughly explain how to take apart, check, and reassemble eleven representative magnetos. If you have an old magneto not specifically covered by the book, you will find that there are many similarities and repair will not be all that difficult.

Old tractors' electrical systems can give you fits even though the magneto is repaired. To help you keep your tractor engine purring like a kitten, I examine basic generators, voltage regulators, starters, and wiring.

I hope that you enjoy reading this book as much as I have enjoyed writing it.

Chapter 1

Basic Electricity

There are two problems in trying to explain basic electricity. First, you must explain electricity. In order to do this, atomic theory must be touched upon. So, you show a diagram of an atom and the reader's eyes immediately glaze over. "This book is supposed to be about an old tractor magnetos, not nuclear energy," the reader might think. I'll keep it simple—don't worry—and give you a few simple concepts that will help you repair your tractor.

The second problem is formulas. Every attribute of electricity can be calculated by using the proper formula, and dozens of formulas are used to calculate this property or that. If you are going to design an electrical device, it stands to reason that you will be better off if you calculate the values of the components needed before you purchase them. On the other hand, if you want to repair a device, it is a matter of measuring a component and comparing the quantity with a known good device.

The only formulas I use in everyday electrical repairs are found in Ohm's Law. Before we deal with Ohm's Law, we must return to the basics and atomic theory.

Atomic Electricity or Electrons on the Loose

All matter is made of atoms. Atoms are the smallest division of matter which retain the character of a known element. An atom is composed of electrons (negative charge) orbiting around a center (nucleus) composed of neutrons (no electric charge) and protons (positive charge). The positively-charged protons and the negatively-charged electrons are normally found in equal numbers in the atom. For example, hydrogen has one electron, one proton, and one neutron. Oxygen has eight electrons, eight protons, and eight neutrons.

The electrons of some atoms are bound tightly to the nucleus while the electrons of other atoms are more loosely bound. These electrons are known as "free electrons" since they can easily be freed from the nucleus. Free electrons are of interest since they make up an electric current.

If an atom loses one of its electrons (a negatively-charged particle), it becomes positively charged. Since there are now more protons than electrons, the atom becomes positively charged and is known as a positive ion. The atom which gains the loose electron is now negatively charged and is known as a negative ion. In other words, atoms become charged by gaining or losing electrons. Electrons can be moved in three ways—by friction (static electricity), chemical reaction, and magnetic force.

Three Ways to Free an Electron

The first way is by friction, otherwise known as static electricity. If you rub a glass rod with a silk cloth, the glass rod will give up electrons and become positively charged and the silk cloth will become negatively charged. If the glass rod touches something small with a neutral charge, some electrons will transfer to the rod and positively charge the small object. When I was a young lad my science teacher used pith balls suspended from a wire by threads to illustrate static electricity. When the glass rod was brought near one of the balls, the ball was attracted to the rod. When it touched the ball, the rod would attract some of the electrons and leave the ball with a positive charge. When the second ball was touched, the same thing happened. Both balls become positively charged and repelled each other. The teacher then ran a plastic comb through her hair. The comb became negatively charged and when it was moved near the balls, they were strongly attracted. Unlike charges attract each other.

Chemical reactions can impart a charge, as well. Automotive batteries are one example of this. When lead peroxide plates and sponge lead plates are immersed in a dilute solution of sulfuric acid and water, one set of plates becomes positively-charged and the other becomes negative-ly-charged. If you wire a light bulb between the plates, the electrons flow through the filament causing it to glow. The chemical action causes the lead and lead peroxide to turn into lead sulfate; the dilute solution of sulfuric acid turns into water.

The most common source of electrical energy is produced by the interaction of magnetism with conductors. To understand how magnets and conductors generate electricity, you need to be familiar with Ohm's Law, which explains how current (amps), electromotive force (volts), and resistance (ohms) interact.

Ohm's Law

Current is the intensity of flow. It is measured in amperage, or amps, and is represented in formulas by *I*. Electromotive force (EMF) is the attraction of negative to positive. The measure of EMF is voltage, or volt, and is represented in formulas by *V*. Resistance is the measure of how much power (volts/amps) is required to move electricity through a substance. Substances like copper allow electricity to flow readily and have a very low resistance. Things like rubber, which allow little or no electrical flow, have an extremely high resistance. Resistance is measured in ohms and is represented in formulas by *R*.

Amperage, voltage, and resistance are directly related. In order to raise voltage, you must drop amperage. Higher resistance will lower voltage and amperage. Thus, the following formula, known as Ohm's law.

$$I = \frac{V}{R} \quad \text{or} \quad Amps = \frac{Volts}{Ohms}$$

Using this formula, you can calculate one of the values if you have the other two. For example, if you have four volts running through a two ohm resistor, there are two amps of current flowing.

$$Amps = \frac{4 volts}{2 ohms}$$

$$Amps = 4 \div 2$$

$$Amps = 2$$

If you double voltage to eight volts, you have twice the amperage (four amps).

$$Amps = \frac{8 volts}{2 ohms}$$

$$Amps = 8 \div 2$$

$$Amps = 4$$

If you double resistance, you get half the amperage (one amp).

$$Amps = \frac{4 volts}{4 ohms}$$

$$Amps = 4 \div 4$$

$$Amps = 1$$

Ohm's Law can be used to calculate any of the missing values, by transposing the values.

$$Volts = Amps \times Ohms$$

$$Ohms = \frac{Volts}{Amps}$$

With these basic tools in hand, we can move on to the basic building blocks of generating electricity: magnets, conductors, capacitors, and inductors.

Magnets: May the Force Be With You

There are two kinds of magnets: natural and man-made. Natural magnets are called lodestones. For centuries, they were the only source of useful magnetism and were primarily used to manufacture compass needles (if a steel needle is rubbed against a lodestone it will become magnetized—insert the needle in a piece of cork and float the cork in a cup of water and the needle will align itself north and south; you have a workable compass). Man-made magnets fall into two categories: permanent and temporary. Permanent magnets are made of very hard steel or of alloys such as alnico. Alnico magnets are made from an alloy of aluminum, nickel, cobalt, and iron. Although many other magnetic materials have been developed in the last thirty years, we are concerned with the first two: hard steel and alnico. Temporary magnets are mostly made of soft iron or steel. Like electricity, magnetism can be conducted. When a soft steel bar comes into contact with a magnet, a temporary magnet will be created.

A magnet has two poles, a north-seeking pole and a south-seeking pole. If you break a magnet in half, each piece will have a north and a south pole. A magnet emits a magnetic a field, which is usually illustrated by a series of lines (see illustra-

A two ox-power single bottom rig with small assistant. I asked the operator about the advantages and disadvantages of the outfit. "Don't have to go to town for fuel. Models didn't change from year to year," he said. The disadvantage? "You can't park it in the machine shed for the winter. Oxen must be exercised every day in good weather or bad."

tion). If you place a heavy piece of paper or a thin piece of clear plastic over a magnet and scatter iron filings on top of the paper, the filings will follow the invisible lines of force (if you poke a magnet into the residue under your bench grinder (wear gloves) you will most likely find enough filings to do this experiment). If a piece of soft iron contacts a magnet, the filings will show how the lines of force follow the iron, which became a temporary magnet. Wrap a wire around this piece of soft iron and run a current through it and you have an electromagnet, which is another form of temporary magnet. The windings of your magneto are one example of this.

When a conductor moves through a magnetic field, or "cuts" the magnetic lines of force, electro-motive force is generated. If the conductor completes a circuit the induced voltage will cause a current to flow. When current flows in a conductor two things happen; a magnetic field is created and heat is produced. If the conductor is formed into a helix or coil, the magnetic field will be intensified. If a core of iron or steel is placed in the center of the coil, the magnetic field will be further intensified. If another source of electricity is connected to the coil, the resultant current flow will cause a magnetic field to form.

All on Board (Conductors)

For practical purposes, the most common conductors are made of copper. Gold, silver, and platinum actually make the best conductors, but are

A typical condenser with the can opened. Rubber rings keep the foil from touching the sides while the spring keeps the foil against the insulator at the top. This metal "can" type condenser had been around for years. Minor changes in sealing and manufacturing made them very reliable.

very expensive. Even so, gold plating is used on connector points (or contacts) where high quality and long life or use under severe conditions are encountered. During the twenties, thirties, and forties platinum points were used in aircraft magnetos as well as tractor applications. Race car magnetos manufactured by Vertex still use platinum points. Aluminum wire is frequently used in the manufacture of starter field coils. Iron or steel makes up for its poor conductivity by size. In the practical world, copper is the king of electrical conductors.

Insulators

Once you have a good conductor, a good insulator becomes equally important. Dry air is an excellent insulator, as is oil, mica, rubber, thermoplastic, varnish, wax paper, beeswax, glass, cotton, and wood. Rust on steel, tarnish on tungsten, and green slime on copper are also good insulators, and are usually found when and where they are least wanted.

Inductance

When current flow increases through a coil of wire, the expanding magnetic field induces voltage in the opposite direction. This voltage opposes change in the flow of current and is known as inductance. The symbol for inductance is *L* and it is measured in henrys. Before you throw the book up against the wall in disgust, I want you to just breeze through this part of the chapter and make a mental note of the various names. In the repair business, we never deal with henrys in calculations. Understanding inductance and its effect in a circuit is very important; calculating inductance is not required. Whenever you look at a coil of wire, consider the fact that the coil will resist any change in the flow of current.

Capacitors (the condenser)

Another key part of the electrical system is the condenser, which is simply a capacitor. A capacitor consists of two metal plates separated by an insulator. A capacitor stores electricity and discharges it when a circuit is opened. You can make a capacitor by placing two metal plates close to each other, but not touching. If you connect one plate to the positive terminal and the other to the negative terminal of a battery, current will flow as the capacitor gains a charge. Disconnect the battery and touch the leads together and the stored current will flow as the plates discharge.

How can this be? Look at it this way. The positive battery terminal lacks electrons. The negative side has an excess amount. If you connect the two terminals, electrons will flow from the negative to the positive terminal. When the wire is hooked up to the capacitor, the capacitor plates build up a store of electrons. The number of electrons which may be stored on the plates depends upon the size of the plates, and the distance they are apart. When a capacitor is fully charged, an electrostatic field exists between plates. The closer the plates are to each other, the stronger the electrostatic field becomes, and the greater the number of electrons which can be stored. If the field becomes too strong for the air gap or insulation, flash-over may occur. Capacitors are classified by capacitance, 0.25 microfarad for example, and by the maximum voltage they can sustain without flash-over.

In the practical world, except for the variable capacitors used in the tuning circuits of radios, the plates are separated by insulators. Mica-filled Bakelite insulation, when substituted for air, will increase capacitance by a factor of five. Other insulators used are mica, glass, paper, mylar, and oiled or waxed paper. The plates may be made of copper, tin, or aluminum. The capacitors in tractors and automobiles usually use foil plates separated by waxed paper. The plates are typically rolled up and installed in a can type container, with one set of plates fastened to the can, and the other ending in a wire at one end of the can.

Some of the old Edison-Splitdorf magnetos used a capacitor mounted on top of the coil. This condenser sits out in the open with heavy mica insulators top and bottom and brass caps on either end. Heavy straps fasten to ground on one side and the wire to the coil and points on the other. The surprising fact is that these condensers rarely go bad. They don't have much capacitance for their size, but they do the job.

Electrolytic capacitors contain either a fluid or a paste which, along with the negative post, forms the negative electrode. The electrolytic capacitor has considerable capacitance for its small size, mainly due to a very thin coating of oxide on the positive plate called the dielectric. Electrolytic capacitors are polarized and must be installed in accordance with the markings or they will not perform properly and will soon fail. Under no circumstance should an electrolytic capacitor be hooked up to a alternating current (AC). (One time I desired to determine the capacitance of an unmarked electrolytic capacitor. My usual condenser tester went off scale when I attempted to test the capacitor. "No problem," I said to myself, "I'll use my big capacitor tester." My big tester is used to test AC motor start capacitors, which usually are between 100 and 500 microfarads in capacitance. My big tester operates on 110 volt AC. Trust me on this, electrolytic capacitors explode when tested on 110 volt AC. Sounds like a small fire cracker. Throws shrapnel. Don't do it.

Capacitors and inductors change the way current and voltage changes flow. Normally, current and voltage rise at the same time. With a capacitor, current flow leads voltage rise. With an inductor, current flow lags behind voltage rise.

Now we can delve into the nagging question that is in the back of your mind; how do magnetos create a spark?

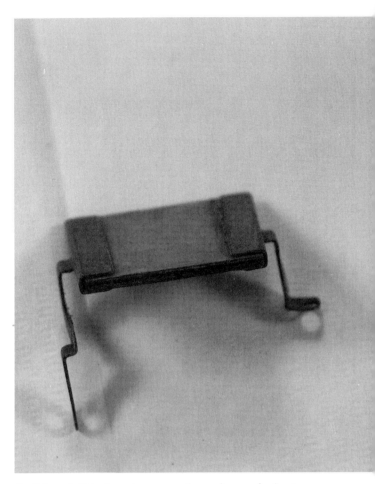

An Edison-Splitdorf condenser uses heavy layers of mica top and bottom and very thin layers of mica between the layers of foil. Brass end pieces hold everything together. These are the most durable and reliable of all the old condensers. I am always surprised when I find one that has failed. This condenser tests at 0.07 microfarads.

Chapter 2

Everything You Wanted to Know About Magnetos, But Were Afraid to Ask

The Random House College Dictionary defines magneto as "a small electric generator, the armature of which rotates in a magnetic field provided by permanent magneto (short for magnetoelectric generator)." We can carry the definition a bit further. Magnetos are self-contained, engine-driven devices that provide a timed and distributed electric spark for the purpose of igniting a gasoline-air mixture in the cylinder or cylinders of a internal combustion engine.

Let's examine how the run-of-mill, battery-powered ignition systems work. The battery-powered ignition system consists of a battery, wire, ignition switch, ignition coil, distributor, and spark plugs. It also contains a set of points (or contacts) connected in parallel with a condenser (or capacitor). Add a cam to open the points at the proper time and we are set for business.

When describing electrical circuits in the real world we can ignore the fact that electrons flow from negative to positive. We will just say that electricity flows from the battery to the ignition switch, and from the ignition switch to the coil. It flows from the coil to the distributor and then to the ignition points. If the points are open, no current will flow. If the points are closed, the current flows through the points to ground. It returns to the battery through the ground circuit.

The coil consists of two windings, one wound on top of the other. The primary winding is a few turns of fairly heavy wire wrapped around the secondary winding, which is many turns of a very fine wire. There is a core of mild steel, usually made up of many laminations riveted together. The secondary winding is connected to the plug wires, while the primary winding hooks to the high-tension cap. In order to complete the circuit, a spark must go to ground through one of the spark plugs and thereby return to its beginning (remember, electricity always flows in a circle).

In the chapter on basic electricity we learned that when electrical current flows through a conductor it creates a magnetic field. When the conductor is wound around a core, a very strong magnetic field is produced. When the electricity starts to flow, the magnetic field induces a voltage in

Magnetos come in all shapes and forms. This one is an internal flywheel magneto from a 1916 Evinrude outboard motor.

opposition to the flow of electricity from the battery (self induction). The expansion of the magnetic field past the secondary winding also causes a voltage to be induced in the secondary winding. This induced voltage is not great enough to cause a spark to jump the gap at the spark plug.

After a short time, the magnetic field is as strong as it can get. The coil is saturated. The points open, and the magnetic field collapses. As the collapsing field moves past or through the coil winding, it induces voltage in both the primary and the secondary coils. Since the primary winding is open at the points, no current can flow in that direction. The voltage in the secondary winding rises to the point at which the spark can jump the gap at the spark plug. The mixture fires. The engine runs.

The Condenser

Of all the electrical devices in the automobile or tractor none is less understood than the simple condenser. The function of the condenser is to insure a clean, non-sparking, opening of the points. Without a condenser in the circuit, electricity would jump the points gap.

The condenser's function can be easily illustrated with a automotive battery, an inductor, a condenser, and some wire jumpers. Attach a jumper from the battery to one end of the inductor (an old ignition coil, an alternator rotor, or the field coil in a generator). Attach a second jumper from the other end of the inductor coil. Attach a third jumper to the other terminal of the battery. Carefully bring the two loose wires together. Pull the wires apart and note the small spark which follows the wires or alligator clips as you open the connection. Next, attach a known good condenser across the connection. Observe how the spark is diminished. The voltage is the same with or without the capacitor (or condenser) in the circuit. I really don't recommend holding on the ends of the wires with bare fingers and performing the experiment. I've done this accidentally and—believe me—the result is shocking.

The condenser has but one purpose: To insure a clean cut-off of current when the points open. Remember, a coil or inductor resists a change in CURRENT, a capacitor or condenser resists a change in VOLTAGE. When the points (or contacts) open, the collapse of the magnetic field causes the voltage to rise. Without a condenser connected across the points, the voltage rise would arc across the points. This would conduct enough current to retard the collapse of the magnetic field. The slower moving magnetic field will not generate enough voltage in the secondary winding to jump the spark plug gap. When the

A Sun distributor test stand. This test stand can spin a distributor-mount magneto up to seven thousand rpm. Most of the time distributor-mount magnetos are used in race cars. The Vertex magneto in the picture is used in an airboat powered by a small block Chevrolet.

points open, the condenser absorbs electrons at a rate which slows the rise in voltage to the point where it is not enough to maintain current flow across the opening points. Once the flow stops, it would require a voltage of several thousand volts to re-establish a flow of electrons.

Check the Condenser

If your tractor magneto has a cover over the breaker plate which can be removed when the tractor is running, you can check how well the condenser is performing by just observing the points. If the tractor is barely running (missing and sputtering) and you observe a large yellow spark at the points, there is a good chance that you have an open or defective condenser. If, on the other hand, the engine is missing and sputtering and you observe "shooting stars" flying all directions from the points, you probably have dirty points. Properly operating points should show either no sparking at the points, or just a tiny blue spark when they open.

The Spark in the Hanger

Once upon a time, I was talking to my friend Willard Steichen, the manager of Southport Airport which was located some miles south of Minneapolis. "You are an electronic genius," he stated. I nodded modestly. "Well, my son is having a terrible time with his motorcycle. He just tuned it up. Now, it only runs on one cylinder." Willard gestered across the hangar floor at his son, who

had just started his Japanese motorcycle (which had a two-cylinder, two-stroke engine).

"That's simple," I said, "he has an open condenser." The kid stopped the engine. His dad shouted across the width of the hanger, "Neil says you have an open condenser." Two minutes later the kid started the bike and it ran perfectly. He rode over to us. "Forgot to hook up one of the condensers." he said. "Well, it could have been defective," I said still rather pleased with myself. Willard was impressed. But it was simple. In the gloom of the hanger I could see the yellow glow of sparking points from thirty yards away.

The difference between magneto and battery ignition is not great. The source of magnetism for a magneto is a permanent magnet while battery ignition uses electromagnetism. Other than that, the components are similar. The points are similar, the coil contains two windings, and a condenser is used to cut off current flow when the points open.

Low base mount Fairbanks-Morse "J" series on left. High base mount American Bosch magneto on right. The difference between the two is 10mm. Low mount is 35mm from base to center of magnetic rotor. High mount is 45mm from base to center of magnetic rotor.

The Make-and-Break Magneto

My favorite example for explaining how magnetos work is the lowly EK Wico which was used on single-cylinder gasoline engines used for stationery power around the farm. This magneto is operated by a clever cam and lever arrangement which pulls a magnetic inductor away from the armature. When the inductor moves it also opens the points. Total movement of the inductor is about 1/4in.

When the engine is not in operation, the inductor rests tightly against two posts. When the operator spins the flywheel, the cam trips the operating rod that in turn rapidly pulls the inductor down, breaking the magnetic circuit. The magnetic field collapses around the coils. Since the points are closed, current is generated in the primary winding. This current creates a magnetic field in opposition to the collapsing magnetic field. The inductor moves another fraction of an inch and the points open. Current stops flowing in the primary winding, the induced magnetic field collapses. Voltage rises in both the primary and secondary winding. The fewer turns of the primary winding prevent the voltage from rising enough to jump the gap of the points. The many turns of the secondary allow the voltage to rise to the point where the spark jumps the spark plug gap and fires the fuel/air mixture. The inductor returns to its closed position. The points close, and we are ready for another cycle.

When the spark occurs, a small current flows through the secondary winding. This causes a reverse magnetic field to be generated. The collapse of the magnetic field is slowed. Voltage decreases and the spark stops jumping the gap. The reverse magnetic field decreases, the magnetic field collapses some more and another spark may jump the gap. If you hook up an oscilloscope to the primary winding and operated the magneto, you see the typical pattern found with battery ignition or magneto ignition.

All of this happens very fast indeed, and there is additional energy stored within the coils and adjacent parts in the form of capacitance. Wherever different voltages come near each other, capacitance exists. Don't let this factor bother you. The work is done by the movement of the magnetic lines of force past the coil wires. Making and breaking the magnetic circuit uses energy provided by the spinning flywheel.

Rip Van Magneto

The really neat thing about the EK Wico magneto is that it can sit patiently year after year doing nothing. The magnetic field, since it is closed, remains strong. The points are held together by spring pressure so that they do not tarnish except

Horizontal flange Fairbanks-Morse used on four-cylinder two-stroke military drone engines. A low-tension coil loads a pair of capacitors which are triggered electronically to fire two spark coils.

after a very long time. If you turn the engine over, the inductor pops open and you get a spark. Embossed in the brass cover is the warranty, "This magneto is guaranteed against defects in materials and workmanship for all time."

A rotary magneto produces spark much as the make-and-break system of the EK magneto does, with a significant difference. All rotary magnetos have a neutral point. When the rotor is in the neutral position there is no magnetic field through the coil. As the rotor turns there is a magnetic field through the coil prior to reaching the neutral point. At the neutral point, the field collapses. When the rotor passes beyond the neutral point, the magnetic field reverses.

There is a neutral point approximately every one hundred-eighty degrees of rotation in a shuttle-wound magneto, rotating inductor magneto, and two-pole rotating magnet magneto. Rotating magnet magnetos can also have four, six, or eight poles. In these magnetos there are respectively four, six, or eight neutral points, and four, six, or eight sparks per rotation. This brings us to a consideration of engine drive ratios.

Engine Drive Ratios

A four stroke (Otto Cycle) gasoline engine requires one spark per two revolutions of the crankshaft. A one-cylinder engine magneto is usually driven at crankshaft speed or a one-to-one ratio. It provides one spark per revolution. One spark fires the mixture at top dead center at the end of the compression stroke. The next spark occurs at top dead center at the end of the exhaust stroke. The second spark is wasted.

Another horizontal flange magneto, this American Bosch semi-low-tension magneto is used on a Climax V-12 natural gas engine. Natural gas engines are used to power compressors used in air conditioning or as stand-by power plants. Each spark plug has its own ignition coil. By eliminating high tension wires, the danger of explosion is reduced.

A two-cylinder engine that has the pistons moving in concert may be fired by a single-cylinder magneto that uses a two-spark coil (a two-spark coil has each end of the secondary winding ending at a spark plug).

A six-cylinder engine usually drives a two-pole rotating magneto at one-and-one-half times the crankshaft speed. This will give three sparks per revolution which the engine requires. The distributor rotor turns one revolution for two revolutions of the engine crankshaft.

I owned an airplane with a nine-cylinder radial engine. The magneto had a four-pole rotating magnet which was driven at one and one-eighth to one ratio. Very confusing!

Matching Magnetos

There are two broad classifications within this definition: internal magnetos (also known as flywheel magnetos) and external magnetos. Internal magnetos live behind the flywheel of an engine and are really an integral part of the engine. External magnetos mount somewhere outside the engine. Replacement of externals is simple—just remove two or four bolts and slip a complete new unit in place of the defective one. This book covers the external magnetos which were used in farm tractors from 1895 to about 1960.

The Society of Automotive Engineers standardized magnetos at an early date. Due to this, magnetos can be divided into a number of categories. There are low-tension magnetos and high-tension magnetos. Magnetos are classified by mode of mounting, that is, flange-mounted and base-mounted. They are further described by direction of rotation into clockwise and counterclockwise magnetos. We also need to know whether or not they have an impulse coupling and the final drive ratio. A final descriptive element is the lag angle (when an impulse coupling is used).

This descriptive information is important when you have to replace a magneto and one with the same part number is not available. If you know all of the above characteristics of your magneto, you can easily replace or repair it.

Cousin Earl and the Earthworm Four

Suppose you have the magneto from your 1928 Earthworm Four (only ninety-nine Earthworm Fours were manufactured. The factory burned down in 1929 and all records were destroyed). The magneto lacks a name plate. It doesn't spark. You wish to acquire a replacement magneto so that you can start up the old Earthworm, but you haven't the foggiest notion of how to find a replacement magneto. The nearest old-fashioned magneto repair shop is several hundred miles away. What to do? You decide to telephone your cousin Earl, a tractor expert.

Earl asks you to describe the magneto. "Well, it came off my Earthworm Four. The magneto's big and rusty and looks old fashioned and must weigh at least ten pounds. It has a big horseshoe magnet. I've never seen anything like it…" you reply. Earl tells you to relax, the Earthworm Tractor Company probably adhered to the standards set forth by the Society of Automotive Engineers (SAE). In order to select a replacement magneto, it is merely necessary climb down the magneto family tree. He gets his chart and starts running down the categories.

Since it is not a flywheel magneto, we can ignore the internal magneto side of the chart. External magnetos are mounted by the base, flange, or distributor. If the magneto mounts on a shelf with four bolts installed through the base, it is a base-mounted magneto. If the magneto has a flange which covers up a hole in the engine, then it must be a flange-mounted magneto. Obviously, a distributor mount magneto must fit into the hole where a distributor would normally be found. (For the individual who wants to replace a magneto with battery ignition, several manufacturers made kits which bolt in place of a flange-mounted magneto.)

Base-mounted magnetos come in two variations determined by the heights of the centerline of the driveshaft: low base at 35mm and high base at 45mm.

Flange-mounted tractor magnetos usually use a standard vertical SAE flange. Starting engines on Caterpillar Tractors sometimes use the horizontal flange as do most light aircraft magnetos. Wisconsin magnetos frequently use the small, round, Wisconsin flange. Occasionally you run

A DU4 base-mounted Bosch magneto. Note the direction of rotation indicated on the oiler cover. The device that looks like a spark plug is the high voltage pickup. One contact goes to the ceramic cover partially visible at the top. The top lead goes through the center of the distributor rotor and contacts the center of the distributor cap.

Driven end of a Wico C-series magneto. The center is six marks from either end and indicates a lag angle of fourteen degrees. Each mark is five degrees apart. The witness mark is left of center which indicates a counter-clockwise magneto. The lag angle is thirty-two degrees. By loosening the four pawl plate retaining screws and moving the pawl stop, this magneto can be made to work on different tractors.

across a non-standard mounting flange on a special application.

After you have determined the mounting type, it is time to determine the direction of rotation. Direction of rotation is usually stamped on the driven end. The term may be "right hand," "left hand," "clockwise," or "counter-clockwise," or "anti-clockwise." If the magneto you are examining has an impulse coupling, try turning the coupling first one way and then the other. The direction that causes the coupling to lock up is the direction of rotation. If the pawl stop is visible and at the lower right hand side, the magneto is clockwise. If the pawl stop is at the lower left hand side, it is counter-clockwise. (Certain magnetos have the pawl stop near the shaft: you can't see it without removing the impulse cup.) Some of the older magnetos have an arrow indicating the direction of rotation.

When there is no impulse coupling and no arrow, turn the magnetic rotor and check for a spark. When you turn the magneto in its direction of rotation, it will produce current and the spark plug will spark. If it doesn't spark in either direction, turn the shaft gently until you can feel the neutral point. The neutral point feels as though a spring loaded ball has dropped into a detent. It resists movement in either direction. Examine the points. They should be closed when the rotor is at the neutral point. The points should open when the rotor shaft is moved about ten degrees in the direction of rotation.

Lag Angle and the Impulse Coupling

We have examined the old Earthworm magneto and classified it as an external magneto with a high base mount and clockwise rotation. Since it uses an impulse coupling, we need to know the

The shuttle-wound coil used in a E4A International Harvester magneto. The condenser is above the coil in this picture. The distributor drive gear is visible as is the collector ring. A brush that protrudes from the bottom of the distributor cap touches the collector ring and conducts the high voltage to the center of the distributor cap where it is distributed by brushes to the spark plug wires.

lag angle. As engine speed increases, the spark must advance, meaning it fires earlier in the compression stroke. To start the engine, the spark must be retarded so that it doesn't kick back when the engine is being cranked. A magneto also has another problem to overcome; at slow cranking speed, the spark is weak.

The impulse coupling solves these problems by retarding the spark and providing the strong spark necessary for starting. The impulse coupling consists of three components: the drive cup with spring, pawl plate with pawls, and the pawl stop or pawl stop plate. The drive cup is connected to the pawl plate by the drive cup spring. It operates

The clever design of the magnetic inductor rotor enable the magnetic flux from the horseshoe magneto to be reversed through the fixed high-tension coil two times per revolution.

An EK Wico magneto in action. Since there was no load, the engine was spinning freely. The magneto would fire only about once in twenty revolutions giving the distinctive "bang—sh—sh—sh—sh—sh—bang." Stationary engine buffs really enjoy it when an engine coasts that way. Under load, of course, the bangs come one after the other and are quite deafening.

in this way. As the engine driveshaft is turned by the starter (electric motor or person with crank), the pawl contacts the pawl stop. This causes the magneto rotor to stop turning. The pawl cup continues to turn, applying torsion the spring. When it has turned a predetermined number of degrees, a protrusion on the cup contacts the pawl and trips it away from the pawl stop. The wound-up spring turns the rotor rapidly and a strong spark occurs. This spark lags the advanced spark position by a number of degrees. The number of degrees difference is known as the lag angle.

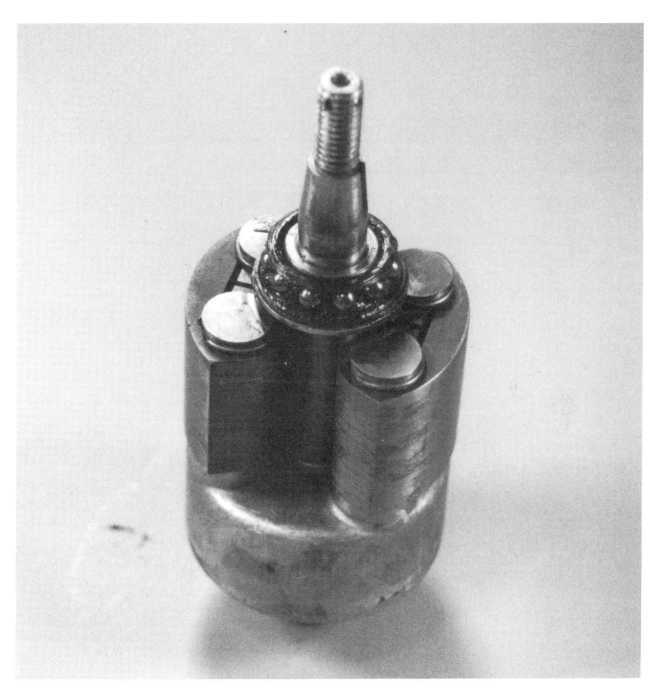

A rotating permanent magnet from a Fairbanks-Morse magneto. I believe this was one of the earliest examples of an alnico magnet cast in aluminum. Magnetism is conducted from the base up the four pins and through the laminations.

The requirement for retarding the spark varies with every engine. If the spark is retarded too much, the engine will not accelerate properly, and the impulse coupling will remain engaged. If the lag is too little, the engine will kick back and probably damage the starter. If you are the starter, you won't like it.

Determining Lag Angle

Since we lack specifications for the Earthworm Four, we can try to find a workable compromise. If the original magneto is somewhat functional, you can probably determine the lag angle by operating it on a commercial test bench. On one of my test benches I have a degree wheel which can be used

A Wico EK magneto with the cover removed. Magnetism is generated by eight bar magnets across the top of the magneto. The condenser lives behind the bar held in place with four screws. The T-shaped device on the magneto completes the magnetic circuit. When the operating lever pulls it down, the magnetic circuit is broken and the magnetic field collapses which induces current in the primary windings. The contacts are closed. The current produces a magnetic field in the opposite direction of the collapsing field. Further movement of the lever causes the T- shaped magnetic conductor to lose any remaining magnetic contact with posts in the center of the coils. The contacts open and the induced magnetic field collapses instantly, causing voltage to rise in the secondary winding until it is strong enough to fire the spark plug.

with a timing light to determine the advanced and retarded position of the spark.

If the original magneto is completely shot or if you lack access to a test bench, you can determine the approximate workable lag angle by trial and error. I would use a Wico magneto with the same mounting, direction of rotation, and drive lugs that would fit the engine drive member. The lag angle of Wico magnetos can be adjusted by setting the pawl stop. When the witness mark on the pawl stop ring is vertical, the lag angle is thirteen degrees. Each mark is five degrees apart. Turning the witness mark in the direction of rotation increases the lag angle. Start at thirty-five degrees and decrease the lag in five or ten degree increments until the engine starts without kick-back and accelerates rapidly. The impulse should almost immediately stop clicking. Normally the pawls retract when the engine speed reaches 250 or 300rpm. The point of full advance will not change when the lag angle is adjusted.

When you have determined the lag angle that seems to work the best, be sure to make a record. When you know the lag angle, it is easier to find a matching magneto.

Matching Drive Lugs

Matching the drive lugs to the engine magneto drive device is the next problem. Slot width is the important thing. I don't know of any easy solution to a mismatch, other than being persistent in searching for an impulse that matches every requirement. On base-mounted magnetos, like the one on the Earthworm Four, there is a floating drive disk between the drive member of the engine and the impulse coupling of the magneto. The purpose of the floating disk is to allow for a slight misalignment of the magneto with the engine drive member. There have been many variations of the floating disks and it might be possible to use the disk to make a mismatched magneto work.

While the external features of magnetos were standardized, the internal system design was left to the whim of the engineer. Over many years the system evolved. Rotary magnetos, the kind used on tractors, started out by using a shuttle-wound coil which rotated within a magnetic field. (The armature and the coil look like a fugitive from the weavers loom. Thus the designation "shuttle-wound.") Rotating the coil and condenser as well as the ignition points, while effective, subjected the coil to the stress of centrifugal force as well as normal heat. The life of the shuttle-wound coil was short. The magnetos were not as reliable as later designs, and were expensive to maintain. Many British motorcycles used shuttle-wound coils in their magnetos until the mid-sixties. In the US, shuttle-wound coils became rare by the late twenties.

Shuttle-wound coil magnetos were replaced by magnetos that used rotating magnetic inductors and fixed coils. The hard-steel horseshoe magnet continued to be used until it was replaced in the thirties by the development of alnico magnetic material. Alnico stands for aluminum, nickel, cobalt, and iron. The British used the alnico magnet material in magnetos where they still used a shuttle-wound coil. In the US, the rotating inductor magneto was replaced by the rotating alnico magnet, and is still used today. I have repaired Robert Bosch (German) magnetos of mid-thirties vintage which used rotating hard steel magnets. These magnetos are fine magnetos indeed, however, the steel rotors are much heavier than the comparable alnico magnet design.

If the external requirements of mounting, direction of rotation, and impulse coupling match are fulfilled, you can usually substitute a shuttle-wound magneto for a rotating alnico magneto, or a rotating magnetic inductor type, with no problem.

Chapter 3

Component Measurement and Testing

The first piece of equipment I reach for when an electrical problem rears its ugly head is my Simpson model 260 multimeter (also known as a volt-ohm meter or VOM). A multimeter measures volts, ohms, and amps.

The multimeter consists of a D'Arsonval movement, a face containing several scales, and an indicating needle. The movement uses a powerful alnico magnet and a moving coil connected to the indicating needle by an axle. A pair of adjustable hairsprings hold the needle at zero. Remember the old rule: electrical current flowing in a conductor creates heat and a magnetic field. Current flowing in the moveable coil creates a magnetic field, which by design, opposes the field of the alnico magnet. The coil rotates on its axle. When the coil turns, the attached needle moves across the face. The coil turns against pressure of the hairsprings until the two forces balance. Depending upon the setting of the main control, it shows volts, ohms, or milliamps. Remember, it is the flow of a small electrical current that causes the meter to show a value. In measuring voltage and amperage the meter takes its operating current from the circuit being measured (when measuring amps, this small current is returned to the circuit—it does consume a small amount of energy). In measuring resistance, either the one-and-one-half volt or the nine-volt battery provides the current. The two batteries are contained in the meter box.

The current required to cause the needle to deflect full scale may be as small as one milliamp (one thousandth of an amp). By inserting various resistors in series with the meter, a range of voltages may be measured. Resistance and current flow are known amounts generated by the meter, so the voltage may be calculated. The meter's resistors are connected in the circuit or removed from the circuit by turning the main control switch to point to the various scales.

To measure current flow, a resistor is placed in parallel with the meter movement. Again, different value resistors are used to establish different ranges. For example, the scales may read five hundred milliamps, two hundred milliamps, fifty milliamps, or ten milliamps. A separate pair of test lead sockets connects to a zero- to ten-amp direct current scale on the Simpson model 260.

Measuring Resistance

By using a self-contained battery (known voltage) and current flow (indicated by meter deflection), resistance is calculated. The resistance scale is geometric, that is the divisions are not equal in width. Zero to one ohms resistance may require the needle to move ten degrees while the measurement from five hundred ohms to infinite resistance may be only four or five degrees. The resistance scale on most multimeters is the reverse of the other measurement scales. The voltage and amperage scales read from left to right, the resistance scale reads from right to left.

Most multimeters use two separate batteries to provide voltage for use in calculating resistance. One battery is a one-and-one-half-volt C or D flashlight battery. The second battery is a common nine-volt transistor radio battery. To adapt to test leads of different length and resistance, a potentiometer is used to "zero" the ohms scale when the test leads are touched together. The knob labeled "zero ohms" is used for this purpose.

Multimeter scales are given in multiples, for example Rx1, Rx10, Rx100, Rx1000, and Rx10,000 ohms. Set the control knob to Rx1 and read the resistance directly from the scale. Set the control knob to Rx10 and multiply the indicated number by ten. Follow the same procedure with the other numbers (Rx10,000 means multiply the number shown by the needle by 10,000).

Some very inexpensive meters do not have an Rx1 setting. Rx1K is the lowest scale. The "K" at the end means "multiply by one thousand." I don't recommend this type multimeter for use on tractor magnetos and electrical systems or anything else for that matter. This meter's resistance scale can only be used to show continuity. An Rx1 scale can be used to measure the resistance of the coil primary circuit in the magneto, contact resistance, or the field coil and armature resistance in the generator or the starter.

Measuring Current

In measuring current, the meter measures the voltage drop across a resistor and translates this into milliamp readings. In measuring AC (alternating current) the current going to the meter movement must be rectified (changed into direct

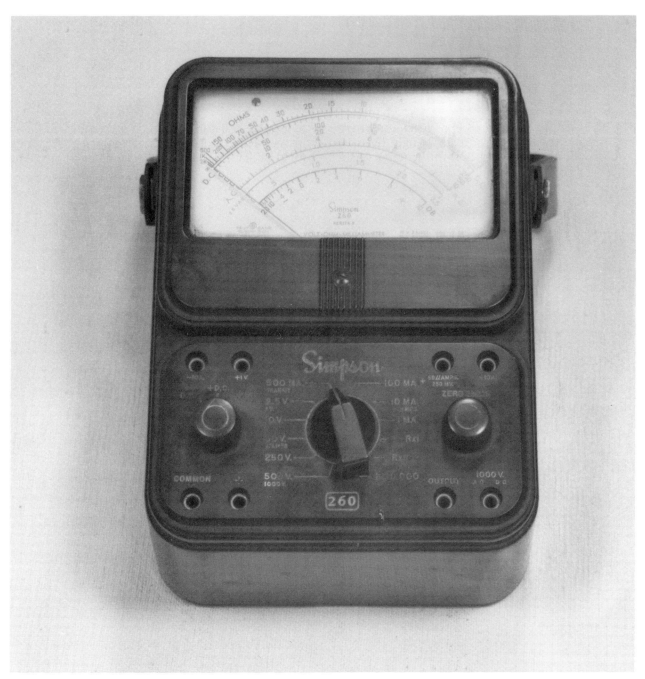

The Simpson Model 260 has been the standard multimeter (or VOM) for most of the last forty years. Most textbooks on electricity or electronics use it as the example of a multimeter. It is rugged, durable, and long-lived. After careful usage over

the last twenty years, I finally managed to damage the RX1 setting of the ohmmeter. I have several back-up meters but I will have this one repaired.

current). This requirement holds true for all AC settings on the multimeter.

Be very careful when trying to measure current. If you accidentally ground a lead instead of attaching it in series to the load, you will blow a fuse (at best) or toast the meter (at worst).

Remember that the meter movement itself deflects full scale with a very small amount of current. In the circuit with the meter movement, resistors reduce the current to a safe quantity.

A military version of the standard multimeter. The case is rugged and waterproof. The meter is high quality. For testing voltages, there are two rows of sockets; the row on the left is rated at 20,000 ohms per volt and the row on the right is rated at 1000 ohms per volt. Why they have two different input sensitivities in the same meter assembly, I don't know.

The range setting inserts the proper amount of resistance in the circuit. If the voltage to be measured is less than fifty volts, but more than the next lower scale, for example, the ten volt scale, set the range to the fifty volt scale and read the quantity on the appropriate scale, which is a scale that ends at fifty, or five or five hundred. When in doubt as to the amount of voltage in the circuit that you wish to test, start at the highest range and work your way down until you get a satisfactory reading. Common sense tells you that, if you had the meter set for the one volt scale, but the actual voltage was, say one hundred volts, the meter would be overloaded. While most meters are protected against overload damage, a severe overload could damage the meter or a component.

A standard multimeter has limitations when it comes to measuring certain circuits. For example, the small current supplied by the internal ohmmeter battery to measure resistance can burn out the delicate sensor of some modern automotive systems. Also, some circuits may cause the meter to give a false reading since the meter uses current from the circuit to operate.

Multimeters are specified by a ohms per volt resistance rating. A cheap meter may have a two thousand ohms per volt rating. A more expensive meter, rated at twenty thousand ohms per volt, would have a resistance of two hundred thousand ohms. As a practical matter, if you are going to use the meter only on old tractor magnetos or electrical systems, an inexpensive meter (with an Rx1 resistance scale) will be satisfactory.

Digital Meters

Historically, the vacuum tube multimeter (VTMM) and the vacuum tube volt meter (VTVM) were developed for very sensitive electronic circuits in radio, television, and audio equipment. Essentially, the VTVM uses a very high resistance voltage divider to reduce the electricity used to make the measurement. The drawback of a VTVM or VTMM is that it requires an external source of power to do its magic. Early vacuum tubes required a substantial current to heat the filament that provides the electrons within the meter, so they usually used 110-volt AC. These analog meters were extremely sensitive, but not very portable.

The development of the transistor made possible more portable, very sensitive high-impedance meters, such as the Sencore Field Effect Multimeter. This meter amplifies the signal present in the same way that the older VTVM did, but can run on batteries.

The more recent development of digital technology and miniature integrated circuit chips

(ICs) has changed the measurement circuitry completely. Digital meters use very high resistance input voltage dividers. The resulting signal is processed and converted from an analog signal to a digital signal by an integrated circuit chip and displayed as a number. The digital meter, like a computer, uses a quartz crystal to provide an internal clock to control a sequence of actions.

A digital meter uses the battery to provide both the operating power for the meter and to provide a reference voltage for measurement. A quartz crystal provides a clock signal. A ramp circuit provides a voltage which increases, within limits, at a constant rate. The unknown voltage, fluffed, buffed, and proportionally reduced by a high-impedance voltage divider, is sent to a comparator. The comparator compares the unknown voltage with the voltage from the ramp circuit. When the two are equal, it stops the clock circuit. The number of ticks which occurred from the start of the cycle until it stops is in proportion to the unknown voltage. The display driver translates this number to voltage and displays it as a number. The voltage, amps, and resistance scale settings control the position of the decimal point.

The display commonly used in digital meters is described as three and one-half digits. The one-half is the first digit, which can be either a zero or a one. The maximum number that can be displayed is 1,999. The smallest is 0.001. The highest number in the 20 volt scale is 19.99. The hundred-volt scale would show the same voltage as 019.9. On the 2000-volt scale, it would be 0020. Four and one-half digit meters are more expensive but are more accurate.

The only disadvantage I have found in digital meters occurs when the circuit is "dirty," meaning it contains rapidly fluctuating voltages or strong magnetic interference. In these cases the numbers fluctuate all over the place, usually so rapidly that reading of the display impossible. If moving the meter doesn't cure the problem, I usually reach for an analog meter. By its nature, an analog meter dampens the fluctuations so that I can read it and get a ball park figure.

The real advantage to a digital meter, besides showing precise numbers, is the additional bells and whistles to be found in reasonably priced digital multimeter. I recently purchased a Beckman meter, which besides the volts, ohm, milliamps, and amps, also had a diode tester, a transistor tester, a logic circuit tester, a capacitance meter, and an audible tone for checking continuity. It also remembers to shut itself off in case I put it down without doing so. For a few dollars more, I could have had a meter that also measured signal frequency and temperature.

Tips for using the Multimeter.

1. When you don't know the voltage or amperage in a circuit you wish to examine, always start at the highest scale and work your way down.
2. When measuring resistance always be sure that power to the circuit has been shut off.
3. Measure voltage with the circuit under normal load. When the circuit is not in use, the meter will usually indicate system voltage. When the component is turned on, a larger than normal voltage drop may indicate a bad connection or a higher than normal voltage draw.
4. Never try to measure high voltages (over 600 volts) unless you are specifically prepared to do so (with a high voltage probe and thorough knowledge regarding its use.)

WARNING: It only takes about 250 milliamps to electrocute a person. This is roughly the amount of energy it takes to light a twenty-five watt light bulb. Be careful!

Coil Testers

Coils are tested by measuring the amount of pulsed direct current required to fire a test gap. If a coil tester says that a coil is bad, the coil is bad. If a coil tester says that a coil is good, it may be good or bad. To be sure a coil is good, it must be tested hot. An early model of the Graham tester had a provision for heating the coil but this feature has not been used since the mid-fifties. Recently I had several questionable coils I wished to test to a dead certainty. The thermostat on my oven had failed and a replacement was not available, so I decided to devise my own system of heating the primary winding with AC power.

By using a large Variac transformer, an AC ammeter, and a timer, I hoped to come up with a reliable system.

It was not. After I had exploded two or three coils in short order, I decided to return to the old oven system. The best way to catch coils that test perfectly cold, but fail under use is to find out if the engine ran fine cold but failed when hot. I do this by asking the owner about the magneto's failure.

When I used my oven, I would heat the coils at 170 degrees for twenty minutes. I would compare the hot performance figures with the cold performance figures. If the comparison showed more than a 0.2 or 0.3 ampere decrease in performance, I would raise the oven temperature another ten degrees and try again. While this system usually caught the borderline coils, the time involved was too great for efficient operation. As an alternative to using an oven, I use my old Eisemann coil tester. I set the control for about fifty percent over minimum test and let it run for a twenty minutes. It the coil still operates properly at the specified test setting, I give it a passing mark. I always take the time to thoroughly test coils where I don't know enough about the circumstances of failure and where I cannot find any other defective parts.

You may well ask, "Why bother with a coil tester, anyhow?" Well, the coil tester does catch most failed coils. The small percentage of coils that fail only when hot can be caught by finding the situation when the magneto quit working.

The Graham Model 41

My all-time favorite coil tester is the Graham Model 41 that I have used for the past thirty years. With the tester set at a specified setting the indica-

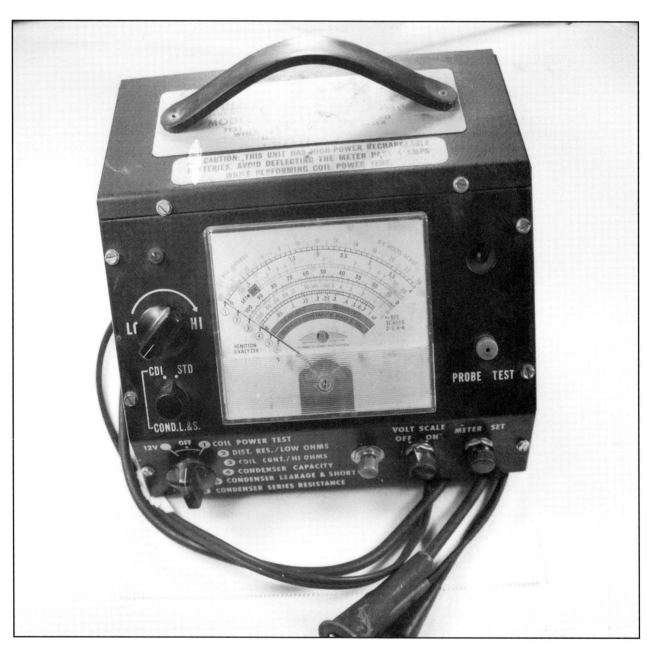

Merc-O-Tronic model 999. This tester measures the amount of current needed to fire a given coil. It incorporates a 5mm test gap and has a prod to test for coil leakage. There are specifications for many coils in the manual. If you haven't got any specifications for a coil, you can compare the amperage re-quired to fire a known good coil with a suspect coil. This tester contains a rechargeable battery to operate the tester and is the unit I take with me when I go to threshing shows. The low resistance ohmmeter section of this tester is almost worth the cost of the whole unit.

tor must read a minimum quantity on the scale. If it fails to reach the proper indication, the coil is defective. If it passes that test, attach a 5mm test gap to the secondary output by an insulated test lead. Reject the coil if it fails to jump the gap. If the spark jumps the gap properly, you go to the last test. Set the control to maximum, attach the test probe to a primary test lead and, with power turned on, run the probe closely over the exterior of the coil. Watch for sparks. Reject the coil if sparks jump through the insulation.

The Graham Model 41 tester also measures capacitance. The more expensive models also measures DC voltage, ohms, capacitance, leakage, and series resistance.

The Merc-O-Tronic

The Merc-O-Tronic Tester, Model 999, also measures voltage, resistance (high and low scale), coil function, capacitance, condenser leakage, and series resistance.

Coils are tested by measuring the amount of pulsed direct current required to fire a test gap. An internal battery powers the tester. Dry cells powered older units. My tester is powered by a rechargeable internal battery. Additional power is available by attaching separate leads to a twelve-volt battery.

The internal battery makes the Merc-O-Tronic tester very portable. The low resistance ohmmeter is very handy for testing magneto coil primary resistance or the contact resistance. The high resistance scale shows ohms times 1000 and an arbitrary scale that is used with the test specification manual. The specification may call for secondary resistance falling between thirty and fifty on the scale, rather than between eight and nine thousand ohms.

The main control knob has six positions. If you start with number one and go to number two through six you will have thoroughly tested the coil and condenser. Remove the coil from the magneto prior to testing.

Spark Test

In the first step attach the positive and negative lead to the coil primary and ground. Attach the high tension lead to the high tension terminal on the coil. With the current control knob turned all the way to the left, move the main knob to position number one. Slowly advance the current control until a spark jumps continuously across the test gap. Note the amperage on the number one scale. Look up the specification for the coil that is being tested. If the spark fails to occur at the test number or lower, reject the coil. If the coil passes this test, go on to the next step.

Leakage Test

Turn the current off and disconnect the high tension lead. Attach the test lead to its socket on the tester and turn the main control to position one. Advance the current knob until it is full on or until the current meter shows approximately four amps. Pass the probe over the surface of the coil while watching for sparks through the insulation. If no sparks jump through the insulation, go on to the next step.

Resistance Test

Turn the main control knob to the low ohms position and measure resistance through the primary winding. It should be as indicated in the manual. Attach the test leads from primary lead to secondary lead and set the control knob on position three. If the meter shows values that are not to specification or an open circuit, reject the coil.

Yes, it is possible for the secondary winding to be open and have the coil pass the first test. However, if a secondary winding is open it will usually fail within a short time. The spark is jumping internally between ends of a wire about as thick as an eyelash. The heat involved will cause the gap to grow until the coil fails totally.

Condenser Test

Advance the main control knob to position four and connect the leads and adjust the meter to the set line. Attach the leads to the condenser and read the capacitance from the number four scale. The proper capacitance can be found in the test manual, if you have the part number.

If you don't have the part number, note the capacitance indicated on the meter. Most modern (1930-1960) magnetos use a condenser of 0.2 to 0.35 microfarads. An antique magneto may be 0.07 to 0.15 microfarads. Check other publications, parts, or repair manuals for an indication of the proper size capacitor. For example, if you have a listing of a magneto that uses the same coil number, note the part number of the condenser and use it.

Turn the knob to position five, turn the small knob from the normal position to the CDI position. Touch the leads together and adjust the meter to the set line with the knob. Connect the leads to the condenser and press the red button. You should hear the vibrator buzz. The needle should momentarily flicker and then return to the small green area. If it doesn't return, but goes into the red area, reject the condenser.

Finally turn the main control to position six, series resistance. Adjust to the set line and connect the leads again. If the needle shows series resistance outside the green area, reject the condenser.

The Eisemann coil tester with an E4 International in place. This tester has been in service since the early forties. I use the Graham or the Merc-o-tronic most often because of the additional measurements they can perform on condensers and coils, but when I have a questionable coil, the final test is performed on the old Eisemann tester. If it passes an extended test, I know that it won't fail in the field.

For years, I didn't check condensers. My theory was, "when in doubt, throw it out." Points and condensers were inexpensive. I replaced both items every time I worked on a magneto. Modern condensers are very durable. Now, I don't replace either unless the part is defective or obviously deteriorated.

Every so often I find a condenser that has the proper capacitance and doesn't leak, but has high series resistance. The magneto barely runs. After I replace the condenser, the magneto runs fine.

If you lack the condenser tester, substitution of a known good condenser is an excellent solution for the problem. Likewise, substituting a known good coil for a questionable coil works. Rare or unusual coils or condensers are hard to find and call for careful testing.

The Eisemann Tester

Magneto manufacturers have often provided coil testers. I have an old tester made by the Eisemann Magneto Company. It uses a standard 5mm spark gap to test the coil. Power is provided by an external battery. The current passes through a rheostat to a set of platinum points, through an ammeter to the coil under test, and back to the battery. The points are opened by a small 115-volt AC electric motor.

The condition of the coil is determined by comparing the current required to fire the test gap with standards for that coil listed in the handbook. When I have a questionable coil to test, I will frequently use the old Eisemann Tester. I increase the current to about fifty percent over the minimum requirement and let coil fire away for twenty or thirty minutes. I recheck at the minimum specification. If the coil still fires consistently, I know it is okay.

The Merc-O-Tronic tester could be used the same way, but I am reluctant to do so. If it dies, the repair will be expensive. If the Eisemann dies, it's no big deal. It is so simple, I am sure I can easily fix it.

The Little Brute

I have a tester made many years ago by the Little Brute Company. It operates on 115-volt AC. It has a dial with two scales, two buttons, and a pair of test leads. To test a coil, attach one lead to the high tension button and the other to the primary lead. Push the test button and read the scale. Compare the reading with a known good coil's indication. If they are close, the coil is good.

If you don't have a known good coil to compare it with, you are out of luck. One of my friendly competitors used a Little Brute for years. He listed the reading of every coil he touched and noted if it ran when he was done repairing the magneto. Eventually, he had a complete listing of coil specifications by part number. This is the procedure used by coil tester manufacturers. Of course, they get the magneto or engine manufacturers to send them sample coils.

Condensers may be tested by connecting the leads across a condenser. Press the test button and read the capacitance on the scale. Release the button and wait a few seconds. Press the button marked discharge. The needle should momentarily indicate full scale. If it does, the condenser is good. If not, test several times before rejecting the condenser.

The Little Brute is not my idea of a precision tester. But, if it says that a coil or a condenser is bad, it is probably bad. Before I got my Merc-O-Tronic tester, I would take the Little Brute with me when I was away from the shop.

The Growler

The growler is a device for testing armatures. It consists of 110-volt coil which lives between the sides of a V block. When an armature is placed upon the V block and the current turned on, the resulting alternating magnetic field reveals shorted windings. To test an armature, place it upon the V block, turn the current on, and hold an old hacksaw blade close to or touching the armature. Turn the armature while holding the hacksaw blade centered. Shorted windings cause the blade to vibrate.

Some growlers incorporate a 110-volt test lamp. It indicates grounded windings. Better quality growlers incorporate an AC ammeter to test output. Test prods contact two adjacent commutator bars ninety degrees from the vertical. Turn the armature, test the adjacent bars, and observe the output. The reading should be about the same all the way around the armature. A zero or very low reading indicates an open circuit.

If you lack a growler with a meter, simply use the hacksaw blade, held vertically, to short circuit commutator bars while turning the armature. The sparking produced as each pair of bars touch the

A growler with an armature in place. If the blade vibrates, the coils are shorted and the armature must be discarded. I always grab the armature next to the commutator and put pressure on the windings as I turn it. If it growls when I squeeze the windings, it will short circuit under centrifugal force.

blade should be about the same. No spark or a very weak spark indicates an open winding.

Open bars can also be detected with a sensitive ohmmeter. If only one winding is open, the meter will still show continuity, but somewhat higher resistance. If two windings are open, you will show infinite resistance across the defective winding.

The growler is not used to test magneto parts, but is still a very useful tool when working on magnetos. When the impulse cup or pawl plate

My state of the art (in 1936) test bench with an F4 International Harvester magneto in place. The bench was a commercial test bench, although I can find no nameplate. The motor is a 220-volt, two horsepower, repulsion-induction motor with moveable brushes. The control knob located between the tachometer and jackshaft controls both direction of rotation and speed. A clever gismo slides on to the driveshaft of the motor with adapters to drive magnetos, gear driven generators, and many other devices. The jackshaft is belt driven and has a variety of pulleys for driving generators and alternators. Several batteries live on the shelf below the bench. These are used for testing generators and motors. They also power the magnetizer which is attached at the other end. Six adjustable spark gaps are not visible in this picture. This is the most versatile test bench I have ever seen. Unfortunately, test benches like this one are no longer made.

touches a magnet it may become magnetized. A magnetized cup will prevent pawls from dropping. A magnetized part may be demagnetized by placing it in the *V* of the growler and turning it on. Gradually remove the part from the growler and move it several feet away, then turn the growler off. Never operate the growler for more than a few seconds without an armature in place. Test the part for magnetism with your compass. If the compass needle is not attracted to the part, it should be okay.

My Test Bench

The piece of test equipment I use the most is my test bench. It was state of the art back in 1936, and is especially useful for checking magnetos that were made around that time. The test bench uses a 2hp repulsion-induction motor with moveable brushes. It runs clockwise or counter-clockwise from 0–3700rpm. When a customer comes to my shop with a magneto that has lost its spark, I will usually install it on the test bench and run a preliminary test before I write up the job. If it doesn't spark, I remove the contact cover plate, or distributor cap, so that I can get at the points. While it is running, I short circuit the points with a screwdriver. If it begins to spark at that time, I allow it to run until the points clean themselves. The customer and I can decide if the magneto should be serviced at this time. If the customer's tractor is down, many times he will take the now working magneto so that he can finish the job. Later he can bring the magneto back for refurbishing. Two out of five magnetos that come to my shop fail to produce spark due to dirty points.

The test bench can also test generators and alternators. Of course, a tractor makes a perfectly good test bench (except in the winter).

Chapter 4

Non-Destructive Dismantling of Old Tractor Magnetos

As there is more than one way to skin a cat, there is more than one way to dismantle an old tractor magneto. I prefer the non-destructive system. The idea is that if you don't break it, you don't have to fix it. There are four requirements for non-destructive dismantling: time, patience, the proper tools, and good technique. If you have a Type A personality—aggressive, hard driving, impatient, a perfectionist—hire someone to do the work. If you are a Type B, or just naturally mellow, continue.

The nemesis of the repairer is the common machine screw. If you can master the removal of stubborn, rusty-headed, narrow-slotted screws, restoration is a piece of cake. If you lack the patience to do the job right, you are doomed.

The engineers who designed those tractor magnetos that we consider to be antique (or at least on the old side) really did a fine job. They used the best materials and technology available at the time. While some modern devices are designed to be thrown away, old machines are designed to be serviced or repaired. Of course, the engineers did not anticipate that, fifty or seventy years after the fact, some fool would try to resurrect and restore a magneto that had been abandoned in the dank corner of an unheated, dirt-floored pole barn since heaven knows when.

The main problem is that the tools that must be used to loosen the old rusty nuts, bolts, and screws just aren't strong enough to do the job. Often, the fastener itself will break, leaving you with the problem of drilling it out and threading the hole. The only thing you can do is to patiently take the time and use the best tool for the job. Use a carefully honed technique, and hope for the best.

Useful Tools of the Trade
Screwdrivers

I never throw any old screwdrivers away and I grab any screwdrivers not tied down. The thing is

you can never have too many screwdrivers. Old mechanical devices used a variety of screws; the secret in removing screws without marring them is to exactly match the blade to the slot. Sure, you can always buy a new set of screwdrivers, but you will still have to modify, sharpen, and shape. With practice and a bench grinder, you can make any old screwdriver do the job.

Examine a new screwdriver carefully. The blade tapers from the shaft to the tip. When a new screwdriver is inserted in the slot, the contact is a rather small area at the top of the slot. The taper works to advantage in this regard. The taper enables even contact to be made in screws with differing slot widths. Also, by tapering to the tip, the screwdriver blade is quite strong. The disadvantage comes with the screw that is very tightly held in place by corrosion. The pressure against the narrow edge of the slot causes the slot to become distorted. The cam effect of the taper also causes the screwdriver to exert pressure upwards and out of the slot.

When you have decided that the screw is very tightly held in place by rust or age, the solution to the problem is to shape an old screwdriver to exactly fit the slot. First, grind the tip by holding it at 90 degrees to the Carborundum wheel. Try to make the tip wider than the slot. Narrow the tip to the proper size by holding the driver at a tangent to the wheel. Exercise patience. When you get it just right, the tip will fill the slot exactly. The sides of the tip will contact fully the sides of the slot. This larger area of contact will enable you to exert maximum pressure upon the screw. The first time you succeed in obtaining a good match the screw will loosen with minimum effort. You will probably decide that you were mistaken in thinking that the screw was tight. Use an ill-fitting driver on the next screw and you will see what I mean.

The engineer who originally designed the magneto, carefully selected the fastener to fit the job. New magnetos were easily disassembled with appropriate tools. Time passes and rust forms. Lubricant that was present when the device was made has long since dried out, adding additional friction where none is needed. After Mr. Rust has moved in, the torque required may exceed the strength of the screwdriver blade. When you turn a common screwdriver very hard, the blade twists. The twisted blade begins to press against the top edge of the slot instead of exerting force against the vertical side of the slot (this action is known as camming out). Naturally, you press down against the screw driver handle and try to keep the blade in the slot. Metal at the top of the slot breaks off or becomes distorted. Less force is expended in turning the screw and more force is required to press the screwdriver into the slot. If you keep it up long enough, you will mangle the screw head

so badly that you must drill it out and tap a new thread.

Finding a matching screw is quite difficult, so be careful so that you don't mar or destroy an otherwise satisfactory antique screw (as collectors know, modern screws on antique tractors are bad—worse than wearing brown shoes and white socks with a tuxedo). When the magneto was first assembled, having an odd screw with a skimpy, narrow slot probably made little difference. When you mangled a screw, you just ordered a new part from the manufacturer.

In the old days they used a variety of slot widths. Some were very narrow, some were extra wide. Occasionally, the bottom of the slot is not flat, but is crescent shaped. It only takes a minute to grind a screwdriver blade to fit. A tight screw, even when you can break it loose, may still turn hard

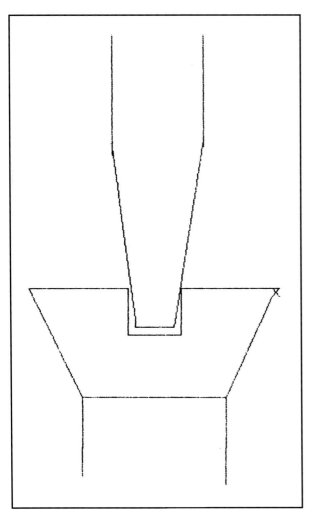

An ill-fitting screwdriver guarantees that you will strip the head and spend some time stressing, straining, drilling, and cussing. The key to extracting stubborn screws is a good fit

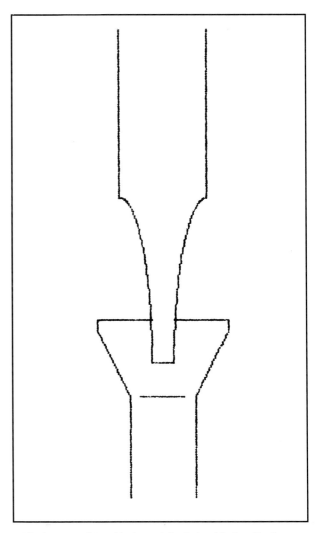

with the screwdriver blade. A tight-fitting blade will allow you to back stuck, rusted screws right out.

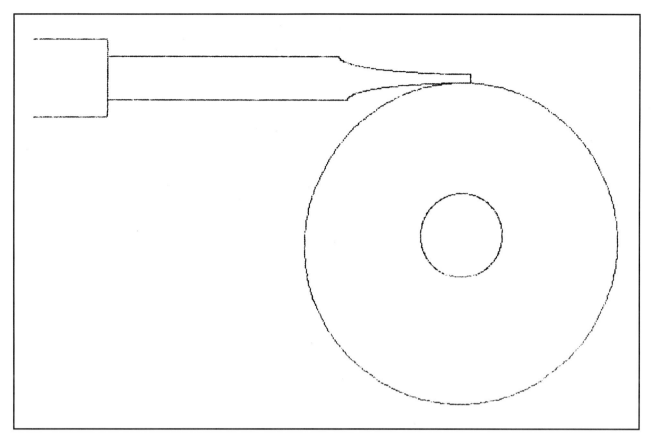

By grinding the blade to the perfect size, the tip of the screw-driver can be custom fit to the screw slot.

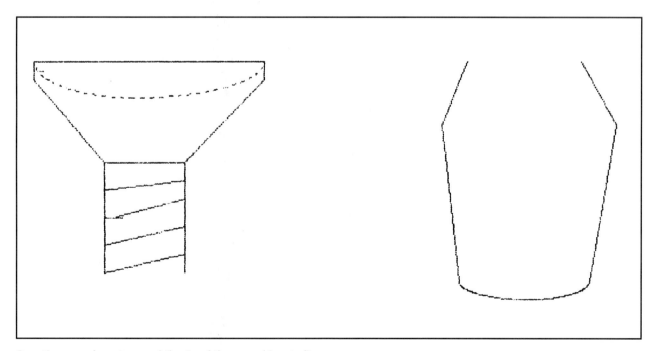

Sometimes you have to round the tip of the screwdriver to fit the slot perfectly.

enough to cause a thin screwdriver blade to become twisted. This will require yet another trip to the grinder. Often screws may be in odd positions inaccessible from straight on. The solution is to grind the blade at an angle. If it works, you can add it to your permanent collection of screw drivers.

Recently, I repaired an old Bosch starter for a customer who sells parts for old European cars.

The end plates were each held in place by six dinky little metric screws. I couldn't budge the screws at the drive end. With a little heat, I managed to remove three screws from the commutator end frame. The third screw cammed out. Violating my own rules, I tried to hurry the repair. That fourth screw didn't want to budge. I was in too much of a hurry to recognize the fact that with

The Dremel Moto-tool is the handiest tool on my workbench. I use the #507 cutoff wheel to trim screws, cut or clean out screwdriver slots, and remove material staked over the seal re- *tainers in Fairbanks-Morse magnetos. The 5/16in fluted ball cutter may be used to prepare a broken screw for drilling out.*

each screw that I had loosened I had gradually twisted the end of my screwdriver blade. I just couldn't get enough bite. The blade had started to "cam out." The same thing had happened to the small bit in the impact driver. Since I was in a hurry to get the job done, I didn't stop then and think a little about what was going on. I just went ahead and banged away until I had butchered the screw slot so badly that I had to stop. I drilled out that screw and threaded the hole. Before I started on the next screw, I sat and finished the last dregs of coffee in my thermos. An examination of the screwdriver under my Sherlock Holmes edition magnifier quickly revealed the blades to be twisted. A quick touch on the bench grinder to straighten the blade and the final screw came out—just like it was supposed to.

I was still in a hurry to get the job done so I slapdashed it back together. It didn't work! In hurrying, I had forgotten one tiny woodruff key, so I had to take everything apart again. By this time common sense had returned. No one was going to die if the starter wasn't done. It was almost eight o'clock on a Friday night, and I was tired. I went home. I did the job right on Monday and the customer was happy.

Dremel Moto-Tool

One of my nephews gave me his old Dremel when he changed jobs. At first, all that I used it for was to clean the recess before installing a seal retainer in a Fairbanks-Morse magneto (the seal retainer is held by staking aluminum over the edge of the retainer). The dremel's cutter was marginally better than a utility knife, but I wasn't too impressed. Later, one of my customers gave me a tube containing part #406 cutting wheels, which changed my outlook. These little cutters make short work of cutting through screws that are too long, or for cutting a screwdriver slot in the top of a broken off bolt. Now I use them for all kinds of things, for example, straightening out damaged screw slots, before I attempt to remove the screw. My second favorite bit is the 5/16in fluted ball cutting tool, which I use to flatten out broken screws before drilling them out.

Drill Press and Good Bits

I recently purchased a new drill index containing both a set of numbered and a set of fractional drills. Being cheap, I bought low-priced bits (as opposed to quality items). You know, the ones that are made in Taiwan or China and are not of the best quality. They do give good coverage. The index box alone is worth the price. The drill bits that I use most frequently tend to have a short but useful life. When I replace these, I replace them on an individual basis with good quality bits. If you

use cutting oil, it prolongs the life of any drill bit. Some years ago, a friend gave me a drill sharpening device. It is a pain to set up, so I pop the dull bits into an old coffee cup until I feel in the mood to sharpen drill bits. Then I set up the sharpener and sharpen all the old drill bits. (A drill press can be substituted with some kind of holder for a 1/4 or 3/8in electric or air drill that will allow you to drill holes accurately.)

A Good Vise

A good vise is like having an extra hand. For years I survived with a small vise with a standard set of vise jaws. When I had a unit I didn't want to scratch, I would wrap it in old rags and not tighten the jaws too tightly. Naturally, many a priceless piece slipped from the jaws and fell to the floor. The fact that my shop has a wooden floor was the only thing that prevented severe damage. I tried some copper slip-over jaws, but they were more of a hindrance than a help. Finally, one of my good customers, a fussy fellow, made me a set of aluminum vise jaws for my new, large vise. This solved the problems of slipping and marring. Old magnetos will bend if you are not careful where you grab them. Grasp base-mounted magnetos by the base. Flange-mounted magnetos can be safely held by griping the bottom ear of the flange. Care is always required.

Bench Grinder

I have a pair of bench grinders. One bench grinder has a medium Carborundum wheel on one side; the other side has a cloth wheel. The second grinder has a medium wire wheel one side and a coarse Carborundum wheel on the other. When used together, the wire wheel and the cloth wheel fluff, buff, and polish the completed job. The grinding wheels are primarily used to sharpen or to form tools such as screwdrivers and chisels into new shapes.

Impact Driver

The impact driver is another GREAT invention. An impact driver is a handy device about 5in long and 1-1/4in in diameter, with either 3/8 or 1/2in square drive shank at one end and a solid cap at the other. You may install either a standard socket or a special socket containing either a common, Phillips, or Torx drive bit on the drive shank. Turn the impact driver in the direction that you wish to turn the screw, hold it tightly and smack the end with a suitable hammer. A clever cam arrangement turns the bit in the desired direction while simultaneously forcing the bit deeper into the screw.

I usually don't use the impact driver on magnetos until I have exhausted the alternatives. On

Before I got aluminum jaws for my vise, I would wrap chromed soft brass or aluminum parts in rags or paper towels to prevent marring. Frequently they would slip out of the jaws and fall to the floor. This was not a desirable way to work. Aluminum jaws are definitely the way to go.

starters and generators (or anything made in Japan that uses Phillips screws), this tool is invaluable. I use mine primarily for removing and replacing starter and generator field coil screws. These screws are usually fairly large and must be tightened firmly indeed. When I must bash on a magneto (after I have tried all the other means of loosening a screw), I carefully grind the tool bit to a hollow ground shape. That way the bit will break before the screw head does. If everything else fails, I will use the standard, blunt screwdriver bit and if the screw breaks, so be it.

Propane Torch

Of necessity, a considerable portion of a magneto is made of non-magnetic materials. In the early days this meant brass or bronze, or both. Bronze has to be about the most durable metal made by man. It laughs at salt water, and withstands the worst weather without flinching and it sneers at the passage of time. I can't remember ever having to drill out a screw from a bronze cast-

ing, except where a steel screw had rusted away. Normal cleaning leaves bronze looking like it just came from the factory.

Unfortunately, after World War I, most magneto makers turned to aluminum, or sometimes a "pot" metal which they passed off as aluminum. The pot metal magnetos are of little consequence today; for the most part, they have long since perished. Where they have survived, they are beyond redemption. When you find a magneto that appears to be made of aluminum, but sheds dark flakes when you rub it with your thumb, you will know that you have stumbled upon the infamous fake aluminum pot metal (a good aluminum casting, when suspended with a wire, sounds a clear bell like note when struck gently; pot metal just makes a dull thud.)

Good aluminum is quite durable, but it pits and corrodes when submerged in water or when water is trapped within a housing. The handy characteristic of aluminum is its high coefficient of expansion. When you heat it with a propane

The impact driver is a necessity for removing Phillips screws from anything made in Japan, where the use of soft screws and thread-locking compound is universal. Pole shoe screws in tractor starters and generators require great force to tighten or loosen. A large forty-eight ounce hammer and an impact driver with the proper bit is answer to tight pole shoe screws. Wear safety glasses when using the impact driver.

torch, the aluminum expands more than steel. Loosening a steel screw in warm aluminum is easy. For example, in removing the coil retaining screws from a Fairbanks-Morse magneto, I always clean the slots in the screw head and gently try to remove them. If they don't turn easily, I light the propane torch and heat the area around the screw while gently exerting pressure with the screwdriver. Ninety-nine times out of one hundred the screw will come out nicely, usually within thirty seconds. When you reinstall the retaining screws, Fairbanks-Morse recommends sealing the tops of the screws with varnish. By heating them with the propane torch the next time, the varnish will soften and again removal is easy.

Even better than a propane torch is an oxy-acetylene torch with a very small tip, especially when it comes to heating a recalcitrant steel nut. I own such an animal, but the local fire department inspector actually broke out in tears when he was inspecting my premises for fire hazards. I hate to see a grown man cry, so I keep the oxyacetylene equipment at home, where it is rarely used. It could be returned to my shop, but I would have to be licensed as a welding shop. It just isn't worth the trouble.

This old arbor press dates from the forties and is one of the handiest devices for pressing bearings and bushings. Later models use a counterweighted handle with a ratchet. When you let go of the handle it moves up and out of the way. I banged my head on my press about twenty years ago. Since then I carefully park the handle out of the way.

Press

For a long time there was an old mechanical arbor press in my shop basement. We had used it in the manufacture of heating elements and had attached a special tool to the ram. After we had stopped making the elements, Dick, my old mechanic, and I had talked about moving it upstairs, but we had never quite got around to doing it. Well, one day Dick retired. I decided to rearrange the workshop, and to do all the repair work myself so that I wouldn't be dependent upon the hired help (all mechanics are prima-donnas and are frequently obnoxious—since I began to do all the mechanical work myself, I have become a prima-donna mechanic and am thoroughly obnoxious). Anyway, the first thing I did was to recruit some help and carry that press upstairs. It had to weigh at least a couple of hundred pounds. I mounted it back to back with my hydraulic press on a steel work table. The steel table is just behind me when I am working at the workbench. If I want to use the press I just have to turn around and there it is.

Today, I would really hate to be without the old arbor press. Nothing is harder than trying to tap in a bearing that wants to tip in its mount. Similarly, when you are removing a bearing it is much easier if you can carefully control the pressure.

Air Wrench

Occasionally it is useful to remove the drive end retaining nut before opening the other end of a magneto. An air wrench enables one to remove a tight nut without having to hold on to the other end of the shaft. A shuttle-wound magneto hasn't much to grab hold of even if you have already removed the distributor cap and points cover. Here is where the air wrench comes in handy. By pressing on the trigger in short bursts, you use the inertia of the coil or the shaft to help loosen the nut. The air wrench works by providing a series of light blows in the direction you wish to turn a nut or bolt. The repeated blows eventually will loosen the tightest nut or bolt.

For many years I disdained the use of the air wrench. I felt comfortable using a speedy handle, and didn't need the expensive luxury of an air wrench. Then someone offered to sell me a well-used air wrench for five or ten dollars. I bit and was hooked. I have a pair of air lines at my bench. One is at full line pressure. I keep it attached to my OSHA-approved blow gun. A pressure regulator with attached lubricator feeds the other line. The second one is set to a moderate pressure for the air wrench. Lower pressure prevents the air wrench from twisting off the smaller screws and bolts. If I want to remove a very tight nut, I merely switch the wrench to the high pressure line and have at it (all my air-powered equipment is fitted with quick-disconnect fittings).

Slip Hammer

I have two slip hammers: One weighs about a pound and a half and has a collar designed to hold a sheet metal screw for use in straightening auto body parts (car thieves use them to remove tumbler assemblies from steering columns). The other is quite a bit larger and screws into a three-jawed puller that pulls gears and pulleys off shafts.

I use the dent puller to remove blind bushings from various housing. To do this, you find a cap screw of a size that will screw into the bushing you wanted removed. If you cut a wedge shaped groove through the threads, the bolt will become self-tapping and will screw more easily into the bushing. Drill and tap a hole centered in the hex for a #10-24 machine screw. Install the screw into

If you use an air wrench for any length of time, you will hate to be without one. Leece-Neville makes an excellent heavy duty alternator. The through bolts are blocked by the fan, so that the pulley must be removed first and installed last. You can wrap an old fan belt around the pulley and lock it in the vise and struggle with a long handle wrench to loosen it with- *out damaging anything important—like your knuckles—or you can use an air wrench with the proper socket. Inertia will keep the rotor stationary if you apply the air in short bursts. It may take a minute or two but the air wrench will loosen the nut without damaging the pulley.*

the dent-puller collar. Screw the cap screw into the bushing. Screw the dent-puller into the cap screw. Secure the housing in the vise so that you can get a good straight pull and vigorously slide the hammer against the stop to extract the bushing.

If you are willing to spend some money, Snap-on, Mac, and other tool suppliers sell blind bushing pullers with matching slip hammers, and u-shaped arbors. These are designed for removing blind bushings. They can also be used to remove the outer race of three piece bearings.

The real secret in pulling and pushing is to pull or push at exactly the required angle. The reason I like using the arbor press for installing or removing bearings or bushings is because it is easy to press squarely with controlled force. Things that appear to be immovable move easily when you exert force in the right direction. The same

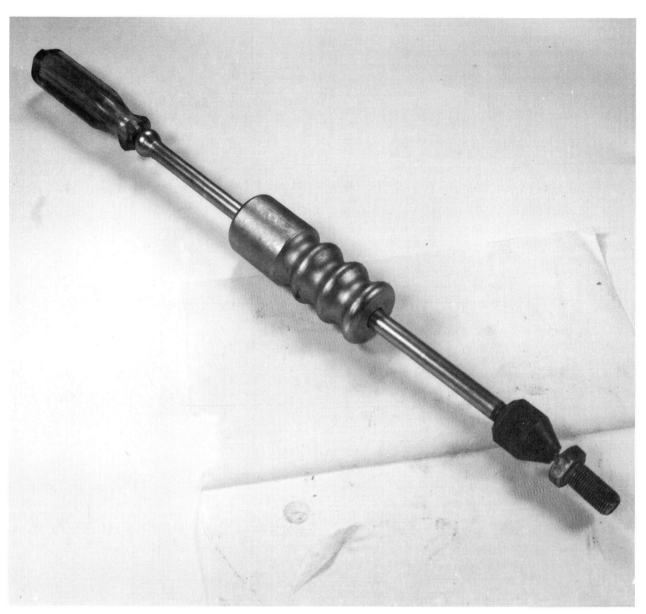

This version of the slide hammer is known as a dent puller. The collar on the end will hold a sheet metal screw. Body shops use them to pull dented metal inaccessible from the back side. They drill a small hole in the panel and screw the sheet metal screw in tightly. Then they slide the "hammer" briskly toward the end of the handle. When the weight hits the handle, the dented metal is drawn out. Thieves use them to remove lock tumblers in steering columns. I attach a bolt of the right size and screw it into a blind bushing. Sharp movement of the weight and the bushing will usually pop out (a blind bushing is one that can only be reached from one side).

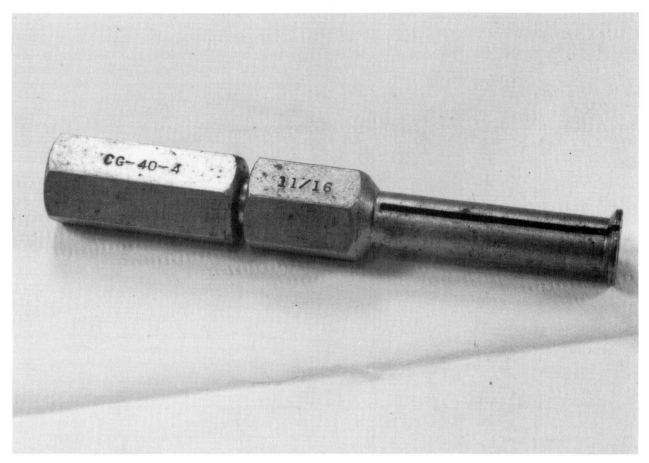

This blind bushing tool is designed for 11/16in holes. The split end is inserted in the hole beyond the end of the bearing or bushing. Hold the split end with an open end wrench and twist the other end with another wrench. A rod pushes a steel ball between the split end forcing the split apart. The ridge on the end of the split is forced past the end of the bushing. A couple of slaps with the slide hammer weight and the bushing will pop right out. I purchased this tool to specifically remove the needle bearing from a Leece-Neville alternator. In this application there is a small hole at the back side of the housing. I was able to press the tool from the back side and remove the bearing and race. This tool is made in sets to fit almost any application.

thing is true when it comes to drilling out broken screws. A drill press will allow you to get precisely aligned. This helps you keep the drill bit from walking from the steel screw to the aluminum body.

Gear & Pulley Puller

I have a whole bunch of pullers but commonly only use three or four. I have several bearing separators that I use with my hydraulic press to remove bearings from shafts. The smaller one works well in pulling the inner bearing race from rotor shafts in magnetos. I have a set of special tools for pulling the outer races from magneto housings.

Pullers can make a fair substitute for a press. Attach a steel plate with a hole in it to a puller and it becomes a "pusher," or a press. When I am away from my shop, I frequently bring my heavy three-jawed puller with a press plate and various odd sized punches along just in case I have to push or pull something.

Penetrating Oils

Penetrating oils tend to be very mysterious. Sometimes they work like magic. Other times, they don't work at all. Everyone always swears by the penetrating oil that worked the last time. Recently I had a customer who was repairing a John Deere engine—that may have been sitting since the last ice age—with water in the cylinders. "I tried everything in the way of penetrating oils," he said. "Finally an old timer looked at that mess and said, 'Water got you into this trouble, maybe water will get you out of it.' So I filled the cylinders with water, and let it sit for a couple of weeks. You know what happened then?" he asked.

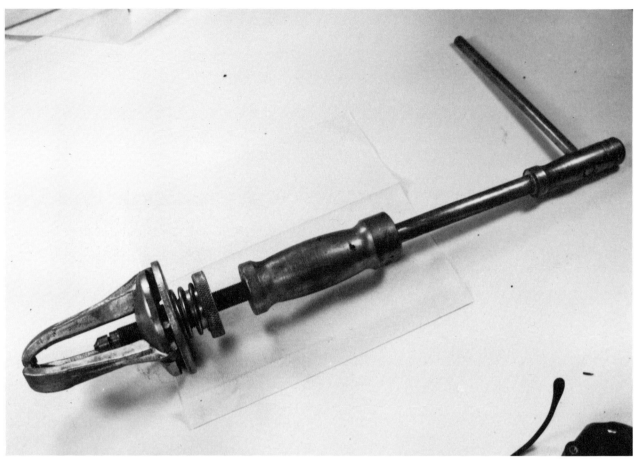

A heavier slide hammer with an attached three-jaw puller. This version can be used as a regular three-jaw puller by turning the shaft with the bar at the top of the tool, or the weight may be slapped against the stop at the handle. A blind bushing puller (see next picture) may be attached to the end of the tool in place of the three-jaw puller.

"You hauled it straight to the scrap yard and got twenty dollars for it." I replied.

"No indeed," he continued, "I nailed a couple of two-by-fours together and pounded them down on the tops of the pistons. It only took a couple of weeks of pounding, and those pistons popped right out."

"Steady pounding?" I asked.

"No," he replied, "I'd try every day or two. Bang away until the pistons moved just an inch or so at a time, and before long they popped right out." (Water as a penetrating oil? I've heard of people who use Coca Cola or muriatic acid. But, I never heard of anyone using plain water.)

A long time ago, somewhere in the early sixties, a manufacturer's representative sold me a case of penetrating oil. Something called Maltby Ferret. We had put it out on the front counter of the parts' department, but we had very few takers. Finally, instead of having it just sit there, I took some back to Roy Valerius, the mechanic in the small engine shop. "Ah, that junk never works" he said.

"Try it anyway," I replied, "the salesman promised that if you dropped just a drop on the rustiest bolt, and let it sit for fifteen minutes, the chances are one hundred to one that it will unscrew itself and you won't have to do any work."

Roy was looking suspiciously at the can when I left. However, when I got back from lunch, he collared me right off, and asked me if I had messed around with that bed knife he was working on. I told him, "No, of course not."

What you should know is that the bed knife of a reel-type lawn mower spends its life being dragged through wet grass and dog crap and trying to cut through baling wire and wire coat hangers and still come up smiling. Eventually they get dull, nicked, and bent. The only good way to sharpen them is to remove the rusted screws or bolts that hold them in place, and sharpen the blade using a special machine. Sometimes when I

An employee destroyed my favorite magneto puller. With great trepidation I called Owatonna Tool Company for a replacement. Thankfully, they had one—the MT8 puller. This puller is invaluable for removing impulse couplings.

would do the occasional sharpening job, I would break off or drill out more screws or bolts than I successfully unscrewed.

Since I was the boss, Roy had dutifully soaked the screws with Maltby Ferret before he went to lunch. When he got back he grabbed his wrench and prepared to do battle. Like the salesman said, they unscrewed themselves. Roy said it darn near killed him when the first screw let go. He was all set to pull hard. When the screw didn't resist, he fell over backwards. Naturally he was suspicious of me, figuring that I had sneaked back and had loosened all of the screws. In this instance I was not guilty. The next time Roy worked on a bed

knife, he sat and stared at the bed knife while the penetrating oil penetrated. The bolts still came out slick as a whistle.

I never saw that salesman again, nor have I seen the product locally. Roy and I grabbed the rest of the cans of Ferret, hid them away, and used them sparingly. They lasted for years. I'll admit to crying a few tears when I used the last drop out of the last can. Alas, I have no more. In all honesty, Ferret didn't work on everything. But when it worked, it really worked.

It seems to me that Ferret, and other penetrating oils, work best when the area around screw or bolt is covered with soft red rust. Sometimes I will leave whatever I'm trying to force apart sitting in the press overnight covered with penetrating oil. The next day, when I have resorted to sheer brute force and forced the recalcitrant pulley from the rotor shaft, I have found that the penetrating oil hadn't penetrated. My guess is that when the expansion of the rusty part is causing the bind, penetrating oil will work. Otherwise, not.

My friend, Frank Bethard, claims that "three-in-one" oil with a dash of ether makes an excel-

lent penetrating oil (use it in a well ventilated place). I haven't tried it yet, but will the next time I get a tough job.

The Right Hammer

When I was fifteen, I filled out an application and sent the State of Minnesota fifty cents and a form and they sent me a driver's license. The next year, when I was sixteen, I took my savings from my newspaper route and purchased a Mod-

This small puller and bearing separator is an easy combination for removing inner bearing races from inductors or shuttle-wound armatures. A larger puller with bolts instead of jaws

works well with some larger parts. There are holes drilled and tapped in the separator jaws for this purpose. The inductor in the picture is from an F4 International Harvester magneto.

I have small, medium, and large bearing separators. I use them to remove gears, sleeves, and bearings from shafts, usually with the aid of my hydraulic press. One time I let a friend use the separator and my hydraulic press to remove a ball bearing from a shaft. Just as he was adding pressure, I noticed that he had the bolts bridging the gap between the support blocks. "STOP," I yelled, " you have the separator crossways. Jeez," I continued, "how can you be so dumb?"

"I'm just following the repair manual," he replied. "Look." Sure enough, the manual showed a picture of someone pressing the identical part with a separator, with the bolts taking the strain.

"Trust me," I said, "let the heavy casting take the strain. I have already replaced the bolts once and once is enough!" I learn from my mistakes.

el A Ford for $200. Two or three years later, while attending the University of Minnesota, I came down with mononucleosis (the kissing disease) and hepatitis. I got over the mononucleosis in a few weeks, but the hepatitis cut down my liver function about one-third. I felt all right and full of energy for about thirty minutes at a time. After thirty minutes I really pooped out and had to go back to bed for a couple of hours, after which I felt full of energy and very bored. So I decided to replace the spring shackles on my Model A Ford. The bushings had long since worn out and the bolts were over halfway worn through. Time to repair.

Well, between the hepatitis and those shackle bolts, I just about died. Every few hours, I would go to the garage and hammer away on those shackles. One pin on each shackle finally came out but the other one would not move. One day Gene Nystrom, a friend who was helping me out, decided to ask his older brother for some help. His brother showed up with an eight-pound sledge hammer, a four-foot steel bar, and an old-fashioned gasoline blow torch. He also brought a punch which was the right size for pushing out the pin and some two-by-fours and four-by fours to brace the spring. We heated the shackle until it was red hot, I held the punch and the steel bar in place and Gene smacked the bar with the hammer. It took about four swats. The recalcitrant pins popped right out. I was astonished. From then on, whenever something didn't fit, I looked for a larger hammer.

This doesn't mean that you should start bashing your magneto with an eight-pound sledge. The point is that you must carefully figure out the amount and kind of force required to achieve the desired result. Too little force is just as bad as too much force. Whereas one heavy blow will usually loosen a tight pulley, for example, many light blows will just form a mushroom on the end of the shaft. Of course, a heavy blow struck off-center will often do real damage. That is why I always use a press where possible.

Impulse couplings, for no apparent reason, will really stick to the tapered shaft of the magnetic rotor. You can tighten the proper puller until the pawl plate starts to bend. No luck? Finally, heat it as hot as you can with the propane torch, chuck the magneto in the vise. Then smack the end of the puller screw with a hammer. Usually, under such stress, the coupling will let go suddenly and you will get to dodge the hot parts. Sometimes nothing seems to work and you have to cut the pawl plate into two pieces (that is destructive dismantling of the old tractor magneto).

The Brass Hammer

If you are going to work around magnetos, a brass or another non-magnetic hammer is essential. The operating instructions for an antique magnetizer recommended striking a magnet with a brass hammer when you magnetized it. The vibration was supposed to help the magnet to become magnetized. So, when the tool sales agent called, I purchased a 16oz brass hammer. One day

a customer wanted to have the magnets on his magneto magnetized. I removed the two thin magnets, which were mounted one on top of the other. To make sure that we obtained the strongest possible magnetism, I walloped that first magnet with my new brass hammer, and immediately turned one large magnet into about one thousand small magnets. I was embarrassed. "Look at the bright side," I said, "When you came in my shop, you only had one magnet. Now you have fifty."

"No problem" he said, "I have plenty of that type to spare." I rarely bang on customers magnets anymore. I do use my brass hammer all the time since it doesn't mark things the way a steel hammer does. Nor does it get magnetized.

Magnifier with Lamp

While it is not as great an invention as sliced bread, my lighted magnifier is the handiest device that lives on my work bench. Whether I am examining identity tags on old equipment or looking for a steel sliver in my thumb, I use it every day. Years ago, my eyes could focus on an object two to three inches away. Now, even with my bifocals on, a foot is about as close as I can focus. This lamp is mounted on an extension arm so that I can pull it into place and it will maintain that position. I can use both hands while I am peeking through the glass.

Some Non-Destructive Dismantling Techniques

The first thing to do is to stop for a minute or two and look at the magneto that you wish to dismantle. Clean off most of the grease, dirt, or loose paint and shake out the leftover wasp nests so that you can see how the magneto is put together. Next, gather the tools that you think you will need. Make sure that your screwdriver blades are the right size. If your memory has started to slip, a notebook and pencil may be helpful On magnetos with horseshoe magnets, I like to make a note about which side the north pole is on. International Harvester frequently stamps an N on the North pole side (a small pocket compass comes in handy). If you forget to do this, don't worry, you can always reverse the magnet if the magneto doesn't perform up to snuff. Most of the time, the orientation of the magnet doesn't make a difference.

Some people like to use a muffin tin when they dismantle a magneto. They place screws and small parts in order in the muffin tin cups. When the time comes to reassemble the magneto, they install parts in reverse order. If they don't have any parts left over when they finish, they can be sure that they did the job right.

My favorite brass hammer. A brass or other non-magnetic hammer is needed to work on magnetos. Brass drifts are also handy since they are not deflected or attracted by magnetism. If you try to tap a freshly magnetized horseshoe magnet into place with a steel hammer you will see what I mean. I have to de-magnetize screwdrivers and pliers on a regular basis, which is trouble enough.

Where to Begin

Common sense tells us to remove the most fragile parts first, and place them where they won't get broken. Having removed the cap and rotor and maybe a coil cover, the next thing to do is to decide which end to start dismantling from. My preference is to start at the driven end and remove the impulse or the drive member. Except for some motorcycle magnetos, magnetos generally have a woodruff key to keep the drive components aligned. Always remove this key before beginning to take the other end apart. If you don't get rid of the key, as surely as the sun will rise tomorrow, that rotor will slide forward and the key will slice a

hole through the seal. It's no big deal. However, a replacement seal is hard to find for some older magnetos. The outer race in the driven end bearing frequently holds the oil seal in place. The driven end bearing is usually the most difficult to reach. If the seal is okay and the outer bearing race is in good shape, it is best to leave well enough alone (if you are going to dunk the whole housing in paint remover or carburetor cleaner it is best to remove both the seal and the outer bearing race first, anyway).

Most of the time I just try to observe the type screw head and length and number that are used, since I usually toss all the parts into a pan containing solvent. Using a small strainer with the solvent pan works nicely. Place the strainer in the pan, and drop the screws as you remove them into the strainer. You can swish them around a bit to clean off the grease and then lay them in a logical order on a paper towel.

I just sort the screws into groups by size, shape and length. Most of the time you find that the plate with four holes will match up with a batch of four screws, a plate or assembly held on with three screws will match up with three similar screws. Of course, when I am dismantling something I make mental or written notes of things that vary from the norm.

For example, the housing which contains the points and condenser in an International Harvester F series magneto, is held in place by two screws and a bolt with an extension on top. The extension holds the metal lever that in turn holds the point cover in place. The two screws are of different length. If you make a mental note that the three screws are different, it is easy to remember what goes where.

Splitting Parts

Along with the worn out screwdrivers, broken knife blades also come in handy for separating aluminum parts. Magnetos that use the three-piece ball bearings usually don't use gaskets between the two major parts of the magneto. Those old-timers did some very precise machining and matching of parts. Frequently, you need to tap a blade into the almost invisible joint between two parts in order to wedge them apart. The aluminum is soft and if you start to pry around with a screwdriver you will surely leave signs behind. Horseshoe magnets that have been in place for sixty or seventy years also require a little persuasion to start moving. Once they have broken loose wiggling and tapping are usually enough to remove them.

Shuttle-wound coils have brass end pieces that bend and distort if handled roughly. They also have a collector ring made, in most cases, of Bakelite (Bakelite, which is still used today, was the first real plastic to be put to common use). These collector rings are not extremely fragile, but if handled roughly will break. To dismantle the armature assembly, first you must remove the inner race of the bearing next to the collector ring. I usually use a special tool for that. I use the bearing separator previously illustrated and press it off using the hydraulic press. You can also use a two-jawed puller instead of the press.

Once rid of the inner race and its spacers, remove the metal shield next to the collector ring. It will slip off easily. Usually the collector ring just slides off. Of course, after being in place for a

long time, you may have to use force. I have yet to devise a tool that will hold the ring without breaking it. So I either curl my left thumb and index finger around the center of the ring and smack the end of the shaft with my brass hammer or I use an underhand hold and slip the ring between my index and my middle finger and again use the hammer. Most of the time it works. If you do break a chunk out of the ring, don't panic. Super glue will come to the rescue. Don't forget to pick up all of the pieces.

I worked on an old Wico magneto that had the slip ring made in three parts, and it screwed on to the armature shaft. I had banged away with the brass hammer, but the slip ring didn't want to come loose. I used the magnifier and discovered something that looked like threads. Joe Malan at Standard Magneto confirmed that the collector ring screwed on. I couldn't unscrew it until I accidentally managed to knock a chip out of the ring. Then I tapped it around with a small hammer. It pays to examine the collector rings carefully.

Drilling Out Broken Screws

The key to drilling out a small broken screw is to drill a small pilot hole exactly in the center of the screw. Over the years I have found one sure way to do the job. First use the Dremel Moto-Tool with the small fluted ball cutter to remove the rough edge of the break. I particularly like the round ball because it enables you to do the job freehand. A flat-end mill of the proper size does a good job, but everything has to be perfectly aligned in a drill press or a milling machine. You can wiggle the round ball until you have a good depression in the center of the broken screw. By removing some material around the edge of the screw, you can see exactly where the edge is. Next, take a needle point punch and tap it into the very center of the screw. I always double check, using the lighted magnifier. If the dimple is right in the center of the screw, I use a slightly larger and blunter punch to enlarge the dent so that a very small drill will be guided to the center. If the small dimple is a little to one side, slant the slightly blunter punch so that the dimple is moved toward the center. When you are convinced that the mark is absolutely on center, use the ball cutter to remove the little ridge pushed up by the punch. This is so that the drill will not be pushed off center.

Mount the magneto housing on the drill press, so that is absolutely level. I usually start with a 1/8in drill bit. Watch when the drill first contacts the broken screw, if it starts to move to one side, carefully align the housing (or whatever you are drilling) so that the drill bit doesn't move off center. Add a drop of cutting oil and drill the hole all the way through the screw. When the first hole is complete, carefully examine to insure that the hole is exactly in the center. If it is, enlarge the hole with a second drill bit a little larger than the first. Check for center again. If it appears to be perfectly centered, select the drill bit that you would use to drill a hole to be tapped to the size and thread of the broken screw.

Drill the final hole. When you are done, you probably can see the edge of the original threads. Use a sharp pick to peel the coil of material that made up the threads of the broken screw. When you do this, the person watching over your shoulder will be impressed. Run the proper tap through the hole to clean up the threads.

If, however, after the first or second drilling you decide that the hole is a little off center, select a drill bit smaller that the one used to drill to tap size. One side of the hole should just touch the aluminum and you should still see in that area the edge of the threads. Take a sharp punch and bend the thin edges of the old screw toward the center. Usually, you can shake the screw out or grab an edge with needle nose pliers and wiggle it out.

If you completely screw up, and have the drill remove all of the thread material on one side, you have two choices. Depending upon the location, and the function of the screw, you can drill out to the next screw size. The new screw will not be properly centered, but it may still work. The second choice is to use something like Loctite Thread Restorer. This is an epoxy formulation that you use to fill the hole before inserting a screw that has been coated with a "release" material. The release material comes as part of the kit. After the epoxy has hardened, you remove the screw leaving a perfect threaded hole behind. I've used this product and it worked well.

Removing Large Broken Bolts

For removing larger broken bolts or studs I follow the recommendations of my good friend Frank Bethard. Frank uses left-handed drill bits (Snap-on Tools and other tool companies have them). First he gets a good dimple dead center. Running the drill press backward and quite slowly, he uses greater than normal downward pressure. If the drill bites tightly enough, it will back the broken screw out. If it doesn't do the job, you haven't lost any time.

For larger bolts, Frank recommends using a drill bushing to center and steady the drill. If the bolt has broken off below the surface of the part, use a bushing with a male thread and screw it into the hole. The bushing will guide the drill all the way through the broken bolt. It eliminates the need to smooth out the broken surface. If the left-handed drill bit doesn't loosen the bolt, you are in good shape to try a straight flute screw extractor (it helps to select a bit of proper size to use a given extractor). Be careful when using screw extractors. A broken screw extractor is very hard to remove. They are too hard to drill: removing one may require drastic and time consuming effort.

If part of the broken bolt is protruding above the surface, a female threaded bushing may be used to make a guide hole for drilling out the bolt or for installing a straight fluted extractor. I almost never use extractors, since the screws and bolts used in magnetos are fairly small and don't lend themselves to extractors. My friend Frank does a lot of automobile restorations where larger bolts are used. He is adamant about not using left-handed spiral tapered extractors. They exert pressure outward against the sides of the stuck bolt which doesn't help in removal. Most of my set of left hand spiral tapered extractors have become permanent residents of the junk pile.

Chapter 5

International Harvester E4A Magneto

The E4A Magneto is a shuttle-wound, base-mounted magneto. It is well designed and very repairable. The most common problem, after dirty points, is coil or condenser failure. If the rotor won't turn easily, the most common problem is insulating tar that leaked from the coil and solidified in the bottom of the magneto. Coils which have leaked tar have usually failed electrically. If a coil feels soft and spongy to the touch, it is in the process of failing, and is doomed.

Farmall Regular with an International Harvester E4A magneto in place. International manufactures almost all of their own components and magnetos are no exception. The E4A and other International magnetos are reliable, high-quality units.

A closer look at the E4A magneto mounted on the Regular.

Unfortunately, the coil is an expensive item to replace. Rewound coils sell for somewhere between fifty and one hundred dollars. I have never been able to find an old, good condenser. You can, however, unsolder the square metal condenser box and install a small capacitor inside. I use an axial lead, mylar insulated, 200–300 volt, 0.1 to 0.2 microfarad capacitor. Use some epoxy or silicone sealer to hold it in place.

Dismantling the E4A

Remove the distributor cap and carefully store it in a safe place. Remove the four screws that hold the magnet retaining strap. Remove the strap and carefully set it to one side. These are often made of brass and tend to be fragile (the brass may be carefully cleaned with solvent and polished with a brass cleaner such as Brasso). Remove the two screws in the side of the magneto housing. One screw is the safety gap screw; it has an

extended point that provides a path for excessive voltage. Excessive voltage may occur if a spark plug wire falls off the spark plug while the engine is running. The second screw is hollow and contains a brush which provides a path for electricity returning to the magneto. Always look for these two screws on any shuttle-wound magneto. Failure to remove the spark gap screw endangers the Bakelite collector ring when you remove the armature. Failure to remove the ground brush will not immediately damage anything. However, if the brush is still in place when you install the armature, it will probably get broken.

The safety gap is found in different places on other shuttle-wound magnetos. Bosch magnetos have the collector ring and pickup at the drive end of the magneto. In these magnetos the safety gap is mounted on the top of the housing and consists of a perforated cylinder 1/2in in diameter and about the same distance high. There is a fine

The distributor cap removed from the E4A. The brush at the bottom of the cover contacts the collector ring and delivers the current to the center brush. The distributor rotor delivers the spark to an outer brush where it is sent to the proper cylinder through the spark plug wire. The use of carbon brushes re-duces arcing within the magneto. Arcing produces ozone. When it mixes with moisture, it produces nitric acid. When you use brushes instead of a jump spark arrangement, ventilation is not needed. I rarely see carbon tracking on the cap or distributor rotor with this system.

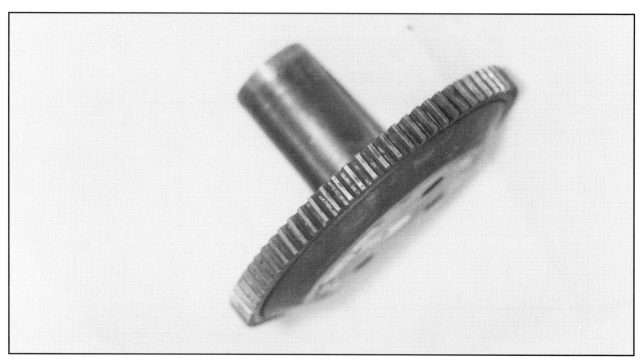

The distributor rotor for an E4A International magneto. There is always a little metal transfer from usage. Eventually, the metal surface wears so that the brushes jump and may arc. I resurface the rotor on a belt sander. Fortunately, the brass contact is quite thick.

mesh copper screen around the inside of the cylinder. The top is a ceramic cover with a contact through the center. Should there be gasoline fumes around the magneto when the safety gap is used, the resultant flame will not penetrate the fine mesh screen to cause a fire.

The pickup brush is located in the distributor cap of the E4A and is kept out of the way of the collector ring when you remove the armature. On other shuttle-wound magnetos the pickup or (in the case of a two spark magneto) the pickups must be removed before the armature is removed.

Remove the Impulse Coupling

Remove the nut that holds the impulse coupling to the armature. Remove the lock washer and the small lipped cup that fits inside of the impulse cup and holds it in place. The impulse cup

An E4A impulse assembly. This impulse assembly was used on many different magnetos provided by Edison-Splitdorf and Bosch as well as International Harvester. The pawl retracts when the engine fires and must be reset before restarting the engine. It is in the set position now.

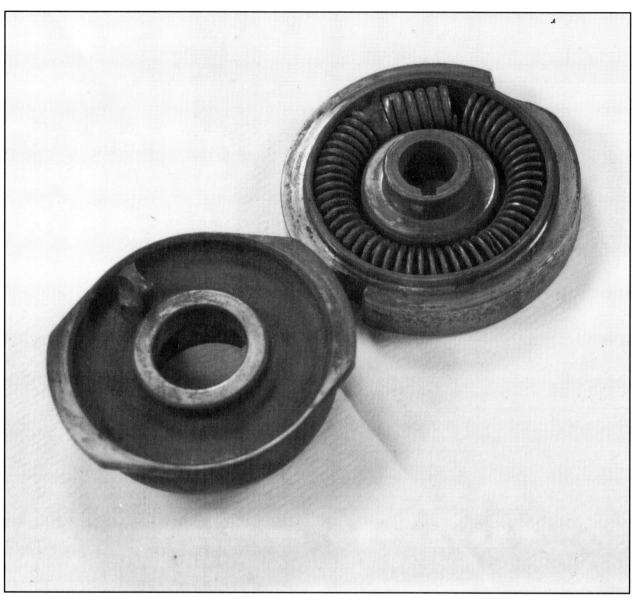

The impulse cup with a coiled spring. The lug in the driven part (lower left) fits between the locator lug in the cup and the end of the coiled spring. When the pawl holds the driven half of the impulse, the coiled spring builds up pressure so that the impulse snaps properly when the pawl is tripped.

may be pulled off by grasping a drive lug with vise-grip or pliers and pulling straight back. Remove the pawl cam plate with a two-jawed puller. If it doesn't want to release easily, heat the plate with a propane torch while maintaining pressure with the puller. A sharp rap on the head of the puller screw will usually loosen the plate.

Remove the impulse spring by slipping the right-angle end of a scribe through one of the turns of the coil spring and pulling the spring free of the channel. The balls or buttons can be removed by sliding them out of the channel at the slightly wider part of the channel directly opposite the stop. There should be a felt string through the center of the spring. I usually clean this in solvent and dry it out. I lubricate the spring and the felt with SAE 20 oil before reassembling.

Remove the two bolts which hold the impulse stop plate in place and remove the impulse stop plate. I usually drop the plate into a pan of solvent. If the magneto has been sitting for some years, the pawl is usually stuck to the pivot; soaking in solvent will usually free it up. Remove the woodruff key from the armature shaft.

Slide the advance plate retainer to one side and remove the advance plate. Unscrew the contact plate retainer screw. If the contact plate doesn't fall out, leave it be for a bit. Remove the three screws that hold the bearing plate in place and gently tap it loose. The contact place will come with the cover and fall out. Slide the armature out. If the coil has leaked tar, the armature may stick in place. I usually place the housing in the arbor press and gently press it out. You can also carefully heat the base to soften the tar and slide it out.

End plate with contacts and cam ring in place. In operation the contact plate with the contacts rotate. The cam ring opens contacts at the right time and a spark occurs. The movable contact is heavier at the rubbing block end. At high rpm, this aids the contacts in closing. I had a customer who complained that his engine was hard to start. He had resorted to pulling the tractor to get it running. Once running, it was strong at full throttle but weak and missing at slow speed. When I took the magneto apart, I found that contact spring was so rusted that it exerted almost no pressure when at rest. When I spun the magneto on the test bench, it had a good spark above 400 rpm. I fluffed and buffed the contacts and installed a good contact spring and it ran with good power at all speeds.

The front view of an E4A rotor. Three screws hold the brown plastic to the distributor gear. If you have the new part, it is easy to make a change. C and A timing marks are at the edge of the brass gear. The rotor has beveled teeth to mesh with the proper mark. C stands for clockwise and A stands for anti-clockwise (counter-clockwise).

Removing and Cleaning the Armature

The distributor gear should slide out with the armature. It floats freely in its bearing, kept in place on one side by the edge of the collector ring and on the other side by the coil end piece. If it doesn't want to move, remove the cap from the other end of the shaft and use a dowel to tap it out. Examine the gear for broken teeth. Examine the bearing journal for scoring or wear. If the bearing is badly worn, a new one can be machined from brass stock.

The brushes which pick up the electrical current from the distributor plate and send it to the spark plug wires cause some wear in the plate. The firing of the spark causes some metal transfer from the brass contact in the disk to the brushes. This causes the brush to bounce at that spot which causes more wear. There are two methods of rectifying the problem. The disk may be chucked in a lathe and the surface turned until it is smooth. A bench sander or a flat surface plate covered with sandpaper can also be used to smooth the surface.

Fortunately, the brass contact is quite thick and a fair amount of material may be removed without affecting the function of the distributor.

Clean the crud off of the armature, paying close attention to the condition of the ball bearings. If the balls are in good condition, clean and examine the outer race for corrosion and checking. Remove the bearing cage carefully and examine the inner race. If the bearings are in good condition, clean all grease off with alcohol, lacquer thinner, or stoddard solvent. Apply fresh ball bearing grease.

E4A housing with the magnet in place. The hard steel magnets made by International are high quality items. A friend of mine owns a 1925 International Harvester Truck. It had been sitting for about twenty-five years when Bob Ripley, my customer, decided to get it running. First thing he did was to bring the magneto to my shop for repair. Before I took it apart, I decided to try it on the test bench. It sparked perfectly from the first revolution. The brass bushing were so dry that they squealed like a banshee. I stopped the bench immediately and cleaned, greased, and oiled it. The magneto worked like a charm.

If the bearings are in bad condition, they should be replaced with new E15 bearings. The inner race can be removed using a special puller designed for the job or by grasping them with a bearing separator and pressing them off in an arbor press. The outer bearings can be pulled by using one of the pullers designed for the purpose. My outer race pullers are badly worn from use over the last seventy-five years. I can rarely insert the puller into place tight enough to press the race out cold. However, the puller which weighs eight or nine ounces will hold tight enough to support its own weight. Careful application of heat using the propane torch will loosen the race so that it will fall freely, propelled by the weight of the puller.

An oil or dust seal is held in place by a retainer under the outer race. Grasp the hot housing, protecting your hand with a shop towel or leather glove. A sharp rap will usually cause the seal and retainer to fall out.

Clean the housing, armature, impulse parts, and so forth to your standard of cleanliness. I usually put the magnet in paint stripper and fluff and buff before I re-magnetize it. If you cannot re-magnetize the magnet, be sure to keep an iron or steel keeper in place. Do not paint the area on the inside of the magnet where it contacts the housing. I grease the area I don't want painted. Wipe the grease off when the paint is dry.

Test the Coil and Condenser

Test the coil and condenser with your favorite testing system. Remember that you must unsolder the coil lead from the condenser in order to get a true test. Using only an ohmmeter, the coil primary should test at about 0.3 ohm. The secondary should test at 6,000 ohms. To test the condenser

A tool for removing the outer race of magneto bearings. Note the splits in the outer edge of this tool. A wedge-shaped screw spreads the tool so that it grips the groove in the outer bearing race. I use the arbor press to push the bearing race from the housing. I find that heat from a propane torch applied to the housing is a big help.

Once the inner race, spacer washers, and collector ring is removed, the distributor drive gear may be pried from its location and all screws removed from the armature. Unsolder the condenser connections and pry the coil loose from each end. Be careful; the brass is easily bent.

with the ohmmeter, use the Rx10,000 scale on an analog meter. The needle should indicate less than infinite reading momentarily when the leads are attached. Just a quick flick with the needle returning to infinite resistance. If the needle shows resistance of less than infinite on a continuous basis, the condenser leaks. If the needle doesn't flick at all, the condenser is open. An open condenser must be discarded. A leaking condenser should be discarded since it is on the way to total failure. If the needle shows very little resistance, the condenser is shorted and must be discarded. If you are in a bind and need the old antique for a show tomorrow, there is a good chance that you can live with a condenser that has a slight electrical leak. Over the long run, the condenser will have to be replaced.

I am a firm believer in testing old coils under load for a considerable length of time. For me, it is a case of survival. If I let a questionable coil slip by, Murphy's Law states that the coil will fail at the worst possible time. Say, when the tractor is one-half way around the loop at a Threshing Show, when I am standing nearby in plain sight. Before I install either a new, or reinstall the old coil in the magneto, I like to put it on my old Eisemann coil tester and test it above minimum test specification for twenty or thirty minutes. If it still passes minimum test specifications at that time, I am sure that it will perform properly in service.

Reassembling the E4A

Press or pull the inner races from each end of the armature shaft. Be careful to remove the shims and spacers and keep them safe. I usually slip them on a piece of tag wire and note on the tag which end they came from. I reuse the shims when installing a new bearing. When the arma-

Fingers are the only safe holder for removing collector rings. The Bakelite is fragile and easily broken. It lives on a soft brass tube which is easily damaged. There is no place to grab with a tool. Replacements are hard to find, but Super Glue Gel does an excellent job of gluing parts back together.

ture is installed I check for excessive end play or tightness.

If the new bearings cause the armature to bind, remove the inner race and several shims. Install the inner race and try again. If there is noticeable play, secure the base of the magneto in the vise. Use a dial indicator with the lever ball against the end of the shaft. Pull and push against the other end and measure the amount of movement. If you are a purist, divide the amount by two and install shims of that thickness on each end (if you are lazy, put all of the shims at one end). This should eliminate the end play. The armature should rotate freely.

In the past thirty years I have not seen evidence of excessive bearing failure in the tractor magnetos that have come through my shop. Some have had a little more play than I felt comfortable with, while others have been snug. Most of the time, bearings are discarded due to corrosion from sitting unused for years and years. Some magnetos have operated for years with, what I would call, excessive end play, showing no ill effects. If they are so loose that the armature rubs against the housing, damage to the armature will occur.

The collector ring takes the high voltage from the coil and transfers it to the high voltage pickup at the bottom of the distributor cap. The Bakelite insulation keeps the spark from jumping to ground.

The inner race puller is used to remove the inner ball bearing race especially at the collector ring end of the armature. I use a large bearing separator to hold the puller in my hydraulic press. Push on the rod protruding from the upper end and the race comes right off. This tool originally used a threaded bolt to do the pushing. After forty or fifty years the threads became stripped. I find that I get good results using the press with my special mandrel.

Grease and Oil

Before installing the armature in the housing, grease the bearings using a good grade of ball bearing grease. Don't overload the bearings with grease but don't skimp either. Most manufacturers recommend filling no more that one half the available space with grease.

Lubricate the wick at the bottom of the distributor rotor housing with SAE 20 oil. Clean the oil passages with air and solvent and if there are wads of cotton in the oil cups, remove them with a tweezers and either clean them thoroughly with solvent or replace them with fresh cotton or felt. I don't like to pack them tightly in place since I want them to pass the oil freely but block out dust. Since you have cleaned and packed the ball bearings with good grease, oiling the passages leading to the ball bearings is not necessary for a long time. When in doubt, however, oil it. If you over-oil your antique magneto, you may have to clean up the excess oil at some point. If you fail to keep oil on a sleeve bearing, or a ball bearing, you get to replace many expensive parts (if you can find them).

A disassembled inner race puller. The split die slips over the inner race and grips the groove. The horseshoe-shaped clip holds the split die in the socket.

Before you try to insert the armature in the housing, make sure that the balls are pressed tightly against the inner race. Match the two beveled gear teeth on the armature with the tooth marked with a dot on the distributor gear. Slide them into place at the same time. You may have to press the oil felt down from the opposite side, so that the distributor shaft will slide into place. Install the bearing housing over the rotor shaft and install the three screws loosely in place. Turn the armature as you slide the housing in place and tighten it down. If the housing doesn't want to slide all the way in place, don't force it.

Try spinning the armature. Some times the balls want to hang up and not slide into the race properly, or the bearing may be pressed to far back on the inner race. Pull everything out and make sure the balls are in place and pressed down in the cage as far as they will go. Most of the time you can feel them click all the way into place and the armature will turn freely as you tighten the screws.

Install the safety gap screw and the ground brush in the housing.

Clean the points of all contaminants and install the point plate in the end of the armature. I

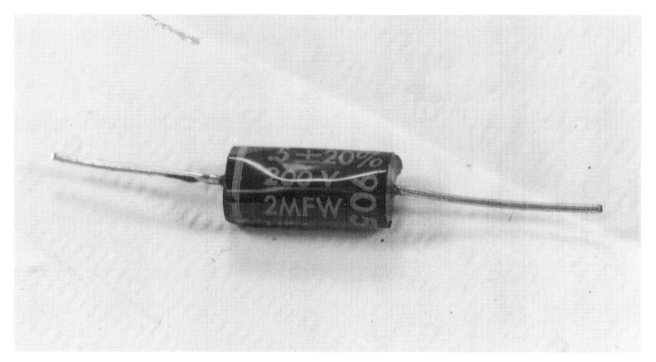

An axial lead condenser, 0.2 microfarad and 200 volt, is all right for most magnetos. It may be installed in an old condenser can of just secured with a drop of epoxy or super glue.

use 0.015in point gap. Install the advance lever and distributor cap then test the magneto. I usually run the magneto for fifteen or twenty minutes on the test bench. After the test, I like to feel the armature for end play and make sure it turns freely. I also make sure that it is not excessively hot. I recommend checking the magneto for ease of rotation, end slop, and excessive heating after it has run for the first time.

A properly repaired and magnetized E4A magneto produces a substantial, impressive spark at a low rpm.

Chapter 6

International Harvester F4 Magneto

The F4 magneto superseded the E4A magneto. They look similar and are interchangeable. Both are base-mounted, clockwise-rotating, four-cylinder magnetos. The F4 uses a magnetic inductor and a fixed coil instead of the rotating shuttle-wound coil. It uses a hard steel horseshoe magnet just like the E4A. Like the E4A magneto, the F4 has a powerful spark at a low rpm.

Dismantling the F4

Start with the distributor cap when you dismantle the F4. Put it carefully away so that it will not be damaged. Remove the two screws that hold the magnet retainer in place, and remove the magnet. Set it on a heavy chunk of steel so that it will retain its magnetic charge. Put the two cap retainers to one side so that you won't lose them. I

A Farmall F-12 with an F4 magneto in place.

The E4 and the F4 can be easily identified from a distance by the way the distributor cap is attached to the magneto. The E4 uses screws and nuts to hold the cap in place while the F4 uses spring clips. From the rear the area below the magnet is open on the E4 and closed with an aluminum cover on the F4. From close up, the model numbers are stamped in the housing below the magnet.

usually stick them on the magnet. When I paint the magnet, I also paint the cap retainers.

Remove the aluminum housing that lives under the magnet. There are two long screws holding the top at the driven end; two short screws come up from the bottom at the cap end.

The Impulse Coupling

The F4 magneto used several variations of the impulse coupling. Sometimes two clips hold a cylinder in place. Remove the two clips and slip the cylinder off. The impulse mechanism uses two pivoting paws which catch on a center-mounted pawl stop. Continued rotation of the impulse cup brings a cam into play which trips the pawl, allowing the inductor to rotate quickly.

Some versions use a thin casting held in place by four screws. Remove this cover. There are two pins that fit into sockets in each part and may remain stuck in either part. The pins provide the pivots for the pawl and pawl trigger. In addition, there is a spring with a loop forced into a recess that maintains pressure on the trigger. The trigger holds the main pawl out of the way when the engine is running. The pawl is pushed up when the engine starts. The trigger captures it and holds it up.

On the back side of the impulse coupling is a pair of secondary pawls. When the engine is being cranked, a right-angle lip on each pawl pops up and touches the trigger assembly and releases the pawl. The pawl falls down and catches the impulse and prevents the shaft from turning. The impulse spring compresses as the cup continues to turn. When the cup has turned about ninety degrees, a cam cut into the edge of the cup forces the pawl out of the groove: The rotor turns rapidly producing a good spark. If the engine doesn't fire, the secondary impulse pawl pops up and hits the trigger and the main pawl falls into the groove and stops rotation of the rotor. The spring compresses, the cam trips the main pawl, the rotor snaps around producing another spark.

When the engine starts running, the impulse pawls retract out of the way and the trigger holds the main pawl clear of the coupling groove. If the engine slows far enough, the impulse pawls pop up and trip the trigger that releases the main pawl. Never allow an engine to run so slowly that the impulse coupling continually engages, as excessive wear will occur.

Remove the coupling bolt, lock washer, and plain washer. There is a set of fine threads in the impulse coupling. When you have the right tool, removing the impulse from the tapered shaft of the rotor is easy. If you are going to work on several magnetos, and if you have access to a lathe, I recommend making the proper tool (my tool was made by Owatonna Tool Company, Part #MT50).

You can remove the coupling with a two-jawed puller. Don't use much pressure as the coupling plate is quite brittle. Heat with the propane torch and try a sharp rap with a small hammer. Remove the cup by grasping a lug with a small vise grip pliers and pulling sharply. Couplings that

With the magnet and the cap out of the way, remove the four screws that hold the aluminum cover and remove the cover.

Remove the two screws that hold the magnet retainer and remove the retainer. I usually use a knife blade to pry the magnet loose at the bottom. Once I have a little space, I use a screwdriver with a wide blade. The aluminum cover is fragile so great care must be used if you pry the magnet at the top.

use a coil spring and two balls come out easily and are easy to install. Some couplings use a "clock" spring that tends to pop loose and unwind unexpectedly when you dismantle it.

Reinstallation of the spring is a matter of engaging the spring in its slot and pressing the two parts together. Increase spring tension by starting to turn the coupling clockwise which increases tension on the spring. To wind the spring, you must slightly separate the cup and the body to clear the mating parts that prevent the spring from unwinding during operation. Once the cup is turned past this obstruction, slip the parts together again. For this operation I have a tool with the proper taper and a woodruff key that I use to hold the impulse coupling while twisting and turning the spring. Once I have completely dismantled the magneto, I like to clean and reassemble the impulse coupling first. Then put it aside

since it will be about the last thing put on the completed magneto (when the rotor is out of the magneto, it makes a convenient tool for reassembly of the impulse).

Removing the Condenser and Points

Remove the woodruff key from the rotor shaft. At the other end, remove the advance lever. Remove the screw that holds the moveable contact spring. Remove the cover retaining bolt. Remove the other two screws that hold the contact plate housing in place. Remove the nut from the condenser and slip the coil and contact wires off. Unscrew the condenser retainer and remove the condenser. Remove the screws that hold the distributor gear support in place. Ease the distributor gear support from the dowel pins. Handle the plastic and wire high-voltage lead mounted on the distributor gear support with extra care as replace-

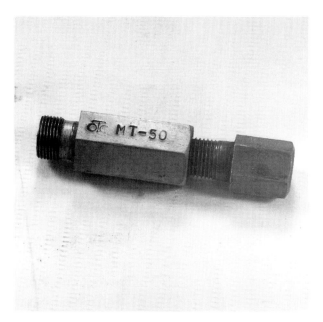

Inspect the bearings for pitting and checking. Inspect the inner and outer races. If everything looks good, clean the bearings in solvent and dry them with compressed air. The IHC magnetos use the E15 bearing, which should be available from a bearing supply house, or a magneto supply shop.

I like to clean all the parts with stoddard solvent, rinse in hot water, and dry with compressed air. Bead blasting the housing improves the appearance. I use round glass beads on the outer housing, magnet, and other parts which leaves a soft, smooth appearance, without eroding the parts.

Owatonna Tool Co. made the MT-50 to remove International Harvester impulse couplings. A piece of hexagonal stock was drilled and tapped. An extension with external threads was cut into one end. A bolt threaded to fit the internal threads of the hexagonal part completes the tool.

ments are hard to find. Remove the distributor gear shims and place them in a safe place. Slip the inductor rotor out. If the lead from the condenser to the points is still in place, swing the housing to one side so that the rotor will clear.

Slide out the metal plate that keeps the condenser to points wire from rubbing on the rotor. Slip the wire from the main housing.

Place the contact housing on top of a partially-opened vise. Use a small screwdriver or brass drift to gently force the contact cup from the housing. Note that there is a pin protruding inside the housing. When the advance lever is moved counter-clockwise, a tab on the moveable contact spring touches the pin and grounds out the coil, shutting off the ignition. Align the pin with the slot in the bottom of the contact cup. Forcing the cup out will allow the point socket to slide out of the insulator.

Early coils have one end of the primary winding soldered to the coil support. A large soldering iron works best to melt this connection: The mass of metal requires considerable heat to melt the solder and release the wire. A number eight flathead screw fastens the soft metal tab, used with later coils, to the coil retainer. Remove the flathead screws on each side of the coil. The coil with its core can now be wiggled free. If you intend to replace the coil, remove the core from the center of the coil. New coils come without the core.

The extension with external threads screws into the internal threads of the coupling. The inside of the hex is large enough to go over the rotor shaft. The bolt screws into the hex and pushes on the end of the rotor shaft, pulling the coupling from the rotor shaft.

Electrical Tests

Measure coil primary and secondary resistance with a multimeter or a digital multimeter. Primary resistance should be about 0.3 or 0.4 ohm. Secondary resistance should be 6000 ohms (plus or minus 1000). Follow the manufacturers directions if you use a coil tester. Discard the coil if it fails any test.

If the magneto failed when hot, test the coil after heating it to operating temperature. If the coil is questionable, once every other possible defect is corrected, test it on the tractor. Test the quality of the spark with the engine cold. Attach a spark test plug to one lead. Make sure the spark will jump at least 5mm (or use an old spark plug with the outer electrode spread to about 3/16in). If you get a good blue cracking spark when the magneto is cold but none when the tractor is hot, I would definitely replace the coil.

Test the condenser with a condenser tester if possible. Lacking a tester, set the multimeter to Rx10,000 and touch the leads to terminal and base. The needle should flick and return to infinite resistance. If it doesn't return to infinite resistance, it is leaking electrically and should be replaced (a digital meter will do the same. Set to the twenty thousand ohm setting, it will show resistance that will increase to the infinite or

Fronts and backs of two impulse couplings. The one on the right uses a clock spring. The coupling on the left uses a coil spring. The short, heavier coil spring is used to change the lag angle.

International Harvester E4A coupling on the left. F4 coupling on the right. Both magnetos use a steel floating disk between the engine drive mechanism and the magneto. The drive disks compensate for minor misalignment.

over range setting). Remember to keep your fingers off the metal part of the prods; the resistance of your body is such that you will get a leakage indication. You can also charge the condenser by connecting the leads across any battery. Wait three or four seconds before touching the lead to the base. You should get a good pop and a spark when you make contact.

In the bad old days, sneaky mechanics would test a condenser by loading it and then handing it to some unsuspecting bystander. They were quite good at measuring condenser quality by observing how high the victim jumped.

The high-tension conductor carries the spark from the top of the coil to the distributor cap. Test

it by measuring resistance between the lead and its mounting bracket. Resistance should be infinite. If you have a source of high voltage like a Model T Ford ignition coil, it is good to test for leakage that way. Every so often you will find a defective conductor.

While you have the high-voltage conductor handy, observe the second wire that ends in space. If you look at the adjacent coil support, you will see a stub wire protruding from the top about 1/8in. The distance between the two conductors makes up a safety gap to handle very high voltage conditions. If a spark plug lead falls off the spark plug while the engine is running, for example, the spark will jump the gap rather than carbon-tracking across the rotor. The gap should be 3/8in or thereabouts. Where the gap is too small, the engine may misfire under load.

Inductor Bearing Replacement

Remove the bearing cages and balls. Remove the inner races with a bearing separator and a press or puller. Save the shims. Remove the outer races. The main housing may have holes drilled in line with the outer race. Use a small punch to push the race out of its housing. Gently, gently! If the race doesn't want to move, try a little heat with a propane torch.

The contact housing usually lacks punch holes, but the outer race is more accessible. I use my worn out race puller, which doesn't put much pressure on the race, and heat the housing. They always come out (knock on wood).

You may find that the bearing has a paper shim around the circumference of the outer race. A paper washer may also be behind the race. I suspect that these insulators were to prevent stray electrical current from passing through the ball bearings. Fairbanks-Morse also used these insulators in the J series magneto. The strips are part #2763. I usually clip a strip off a piece of light cardboard and glue them in place with a thin coat of bearing grease. Lucas, the British automotive electrical system manufacturer, made a washer with ears sticking out all around for their motorcycle magnetos. If you can find some, they really do a good job.

After you install the outer bearing races, the next thing is to install the inner races so that the inductor has minimal end play, but still turns without binding. One way to install the races tightly is without shims. Install the inductor rotor in the housing and install the end housing tightly. Lock the housing in a vise and use a dial indicator to show the amount of end play when pressure is applied first to one end and then the other. Divide the amount by two. Use various thickness shims to make up the proper size packs. Remove the inner races and install the packs to remove most of the end play.

Lazy people just install the shims removed from the old races with the new bearings. If the inductor turns freely and doesn't have excessive end play, let well enough alone.

When I used to rebuild Bendix aircraft magnetos, I followed the manual religiously. Bendix called for installing the inner races and measuring end play very carefully. They also called for having an 0.0005in preload. This seemed to work quite well. Completed magnetos were tested for some hours on the test bench, and they didn't seem to bind. After they were tested, they were dismantled and checked for defects, reassembled

A good view of the trigger pawls the trip the main pawl. Gravity causes the pawl to pivot, extending the trigger. At cranking speed the pawls remain extended, tripping the main pawl at each half revolution. When the engine starts, centrifugal force causes the pawls to retract.

and tested again. I never got any magnetos back. If the planes had crashed, I'm sure someone would have complained.

Assembling the Magneto

Once you are satisfied with the set up of the inductor, install it into the housing. Install the brass fitting with the attached primary lead into the rear of the contact retaining cup. Feed the wire through the bottom of the front housing and slip it up through the notch in the main housing near where the condenser is mounted. Carefully slide the aluminum plate over the top of the wire and insert it in the notches cast into the main housing. Lead the wire up through the groove in the plate. Properly installed, the lead cannot contact the distributor drive gear. Slide the front housing into place over the inductor shaft. Insert the bolt with the retaining arm into the upper left-hand screw hole. Insert and tighten the remaining screws. Double-check the rotor to make sure it isn't binding and the end play is within specifications.

Install the coil, using the number ten flat head machine screws. Put the shims in place. Turn the rotor so that the two beveled gear teeth are at the top. Install the distributor gear with the tooth marked with the dot between the two beveled teeth. Install and tighten the two fillister head

The contact plate end of the magneto. Note the oiler cap visible on the right side of the housing. These magnetos require regular oiling when they are being used. Once a week is about right. Unlike the E4A, the contacts on this magneto are fixed in place and do not spin with the rotor. The contact plate may be swiveled a few degrees by turning the cover so that timing may be advanced, retarded, or turned off.

screws that hold the distributor gear and support in place. Make sure that the fiber gear meshes properly with the steel drive gear.

Install the condenser and bracket and attach the primary lead and the coil primary lead to the stud at the top of the condenser. If the high tension contactor is not in place, install it now. Check the spark safety gap. If the gap is about 3/8in, leave well enough alone. The gap can be adjusted by carefully bending the wire.

Install the ignition points, if you haven't done so. Install the proper screw through the moveable contact clip and tighten. Adjust the point gap to 0.015in.

Before installing the coil cover, check the oil cups and the oil delivery tubes or passages and make sure that oil will pass through. Run a small wire through the tube if it is plugged (there is a little wad of cotton in each passage to prevent dirt from entering). Saturate the oil wick in the distributor gear assembly with a good light (SAE 20) oil. If you have lubricated the ball bearings with good ball bearing grease, they will last for years. If the bushing in the distributor housing runs dry, you

After you remove the two screws and the post with the cover retainer, the bearing housing—which also holds the contacts—comes off the main housing. The piece of metal that prevents the contact wire from getting tangled with the distributor gear will slide out. Remove the condenser retaining screw and the condenser. The two screws that hold the distributor rotor support are removed next. The support can be lifted up from the main body. The adjustment shims may come with the support or remain with the body. In either case, set them to one side so that you don't lose any. One of the screws that holds the coil is now visible.

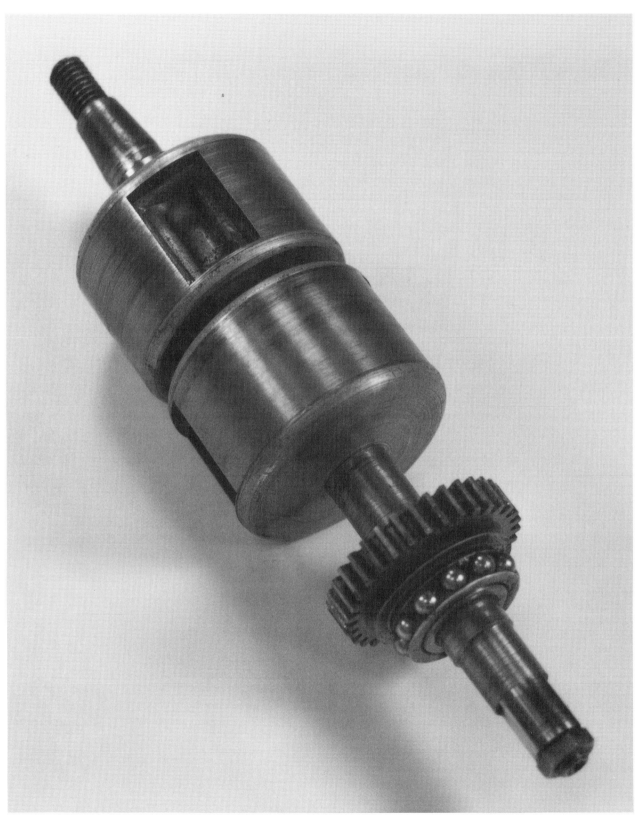

The International Harvester F4 magnetic inductor is made of soft steel so that it conducts magnetism efficiently. Its function is to conduct magnetism through the coil first in one direction, then in the reverse direction. It is held together by non-magnetic material.

International F4 coil with the removed core next to it. After you have removed the coil retaining screws and unsoldered the ground wire, the coil may be gently pried from between the coil supports. If the coil tests bad, the core must be removed. New coils come without the core.

are liable to have a serious problem (broken teeth on the distributor gear, scoring of the shaft, etc.).

Insert the four screws into the housing, slip the gasket in place, install the cover. Install the magnet and the two clips that hold the distributor cap in place.

At this point, I like to spin the inductor rotor with my fingers. If all goes well, spark jumps the safety gap and I feel relieved.

Install the woodruff key in the inductor shaft and install the impulse coupling. Put the pawl and pawl trigger on their shafts and attach the spring to the trigger. Slide the outer cover in place and insert the four screws and tighten them.

I prefer using a vise grip pliers or a drive member to turn the impulse to test its operation. Hold a grounded screwdriver about 3/16in from the high tension brush. If you get a good spark, install the cleaned and refurbished distributor cap and test the magneto on a test bench or tractor.

Chapter 7

International Harvester H4

The H-series International magneto is an excellent, modern, rotating-magnet magneto. The only knock against it is that you have to remove several parts to get to the points. If the points could be exposed with the distributor in place, they could be examined while the engine is running or being cranked. What is happening to the points reveals much of the general health of the magneto.

My old doctor always liked to look at my tongue and throat first thing, whatever my complaint. I had a badly sprained ankle once. First

An International H4 magneto on an International tractor. This is an example of a rotating alnico magnet magneto. Note that it mounts in approximately the same position on the engine as the other International magnetos.

thing he did was to peer into my mouth. I asked him why. He tried to look wise and tapped one finger along one side of his nose. "It's a medical thing," he said. I asked him what I could do with my badly sprained ankle. "Limp," he said. "Now, if you had broken it, we could put you into a walking cast," he continued. "But, with a bad sprain, you just limp."

Well, I always like to look at the points first when I see a sick magneto. My favorite magneto is the MJC-series made by American Bosch. Remove two thumbscrews and a little plate, and you can see exactly what is going on.

Disassembly

With the International H4, unlike the F-and E-series, you must first remove the distributor cap, distributor rotor, and the three screws that hold the distributor gear and support plate in place.

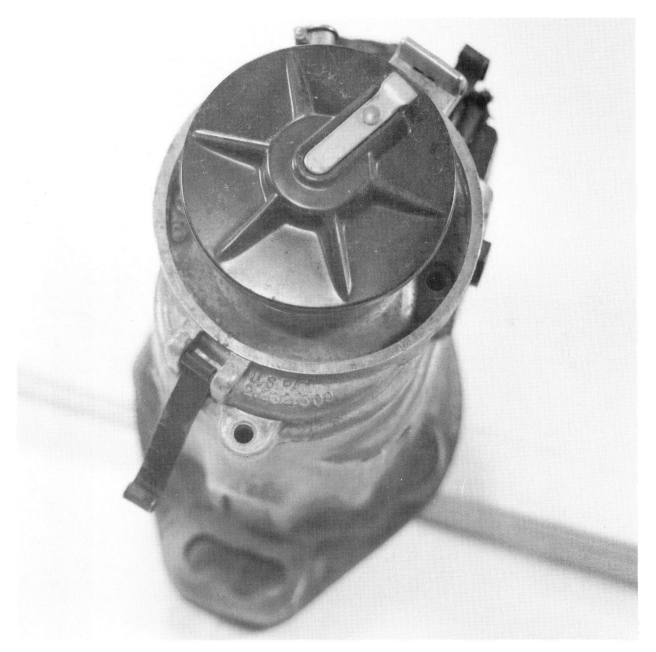

The H4 magneto with the cap removed and put safely away. The rotor is pushed over the distributor shaft.

The condenser lives in its own compartment behind a plate. To remove the condenser, you must first remove the four screws that hold the coil cap to the top of the magneto. There is a long screw with a retainer that clamps the condenser in place. Remove the external shut off stud nut next. The condenser can be shoved though the opening on the other side of the magneto. Carefully pry the coil and contact leads from the condenser stud with a small screwdriver. Note that the coil and the condenser are the same as used on the F-series magnetos.

Remove the two flat head machine screws from the coil and gently tap the coil from its supports. Remove the nut that holds the moveable contact and the connecting wire and slip the

An International H4 magneto with the distributor plate removed. The contacts are easily removed; gently pry the moveable contact off of the pivot. The insulated block also slides out. Note the direction of the insulating block. I always forget and reassemble the insulator backwards, which is very embarrassing if someone is watching over my shoulder.

An H4 magneto with condenser exposed. The condenser re-taining screw is located on the left side. The condenser is the shiny cylinder above the contacts. Remove the retaining screw and the plate on the outside of the magneto. Remove the brass nut on the outside of the magneto at the other end of the condenser. Remove the fiber washer and push the con-denser out the other side. The coil primary wire and the lead to the contacts are attached to the condenser. This setup is a slight pain to put together since the inner insulating washer will fit in only one way, and I always forget how it goes.

The distributor gear and rotor. At the top of the rotor gear, where the rotor fits, a C and A are marked. When you assemble the drive gear to the distributor gear, be sure to line up the marks on the gears. Most, but not all, H4 magnetos are clockwise. When you least expect it, you will find a counter-clockwise H4 magneto.

The new coil has a tab to ground the coil. Note that the new coil lacks the ridges to keep it aligned. The wide tab performs that function.

An old H4 coil with the core partially removed. If you have a bad coil, be sure to save the core. New coils do not come with cores.

moveable contact from its pivot post. Remove the screw that holds the fixed contact in place. I usually reattach the insulation block to the moveable contact with the flathead machine screw and nut. Otherwise, I always install it backwards and have to redo the job two or three times.

Removing the Impulse Coupling

Remove the nut that holds the impulse coupling in place. If you have the right puller, the pawl plate comes right off. A two-jawed puller will work, but be cautious because it is fragile (like all International Harvester pawl plates). Be sure to remove the woodruff key at this time.

Take off the four flat head machine screws that hold the cast iron flange to the magneto body. Save the shake-proof washers on the screws. I always reuse the washers and Loctite the screws when I assemble the magneto. Remove the flange, and the magnetic rotor at this time. Wrap the magnetic rotor in a shop towel to keep it free of metal parts.

Removing Bearings and Seals

The H4 magneto uses the same E15 bearings as the other International Harvester magnetos. Remove and clean the bearing cages. Examine the inner and outer races for corrosion and checking. If the bearing must be replaced, the inner races can easily be removed with a bearing separator. The outer races can be jarred from their homes by using a little heat. I use a well worn set of outer race tools. Mine no longer have the grip to pull the bearing race by brute force, but they will hold their own weight. Add a little heat and just a touch with the brass punch does the job. (Snap-on tool company has a puller that will do the job). A simple leather oil seal is placed in a retainer behind the outer race. If you wish to replace the oil seal, you must remove the inner race and carefully press the seal back through the flange. The seals are easy to find, but the retainers are not. Standard Magneto Sales Company in Chicago, Illinois sells a repair kit for the H4 that contains points, condenser, seal, rotor, and gaskets for a reasonable price.

The same tool used to remove the impulse from the F4 and H4. Be careful if you use a two-jaw puller: The pawl plate is very fragile. Heat and patience will help.

With the impulse removed, the four flange retaining screws and the seal retainer are visible. If you want to replace the seal, you must remove the outer bearing race from the inside of the flange plate first. Note the pawl stop pin on the right hand side, indicating that this is a clockwise magneto.

Reassembly

Wash the parts in your favorite solvent, rinse in hot water, and use compressed air to dry. Test the coil and condenser. Fluff and buff or replace the points.

Grease the ball bearings. Fill the spaces about half full of grease and assemble with the newly magnetized rotor. Install the magnetic rotor in the housing and attach the flange. Install the woodruff key and the cleaned impulse assembly. Install the nut and washer.

Slip the coil into its spot and install the two set screws. Feed the primary lead into place with the contact lead.

The tricky part of installing the condenser is to correctly insert the inner insulator properly in the recess. Slide the condenser home, carefully slip the coil and contact lead terminals over the stud, and push into place. If everything is in place, you can easily insert the screw and retainer. Tighten the retainer screw. Slip the outer insulator over the stud and tighten the special nut.

Install the points and adjust to 0.013in. Attach the coil cover and gasket. Insert the high tension coil lead, mount the magneto in a test bench or vise, and check the spark. If it is good, install the distributor gear and the drive gear, carefully aligning the timing mark on the drive gear with the C marks on the distributor gear. Install the housing cover, distributor rotor, gasket, and distributor cap. Insert the high tension lead in the center tower of the cap. Test on the test bench or install on the tractor.

Chapter 8

The Fairbanks-Morse DRV2 Magneto (John Deere)

This variation of the RV magneto has a horizontal flange for mounting, except it mounts to the John Deere tractor engine in the standard vertical flange position. This leaves the magneto lying on its side. No matter, it works fine.

The impulse coupling appears to be a standard Fairbanks-Morse impulse except for having an ad-

A John Deere Model B with a DRV2 Fairbank-Morse magneto. An old time magneto repairman once told me that the Fairbanks-Morse was a great magneto, "as long as it doesn't rain."

ditional key way. This key way is necessary to make up for the ninety degree turn of the main body. The magnetic rotor is common for all the RV-series magnetos.

Disassembly

I like to start at the driven end of most magnetos because impulse couplings tend to give me the most problems. Once the impulse is out of the way, the rest will be smooth sailing (I hope). If a magneto has a fragile distributor cap, I take it off first and put it carefully away.

Remove the center nut and its lock washer. Use a two-jawed puller and remove the impulse in one piece. If it separates, no problem. It goes together the same way as all Fairbank-Morse magnetos. Remove the woodruff key from the shaft.

Four screws hold the aluminum distributor cap in place. Remove the screws and gently separate the cover from the housing. Examine the two brush-holders for signs of carbon tracking. If you have a high voltage source available, test the holders for leakage to ground. If not, use your multimeter to test for continuity from contact to ground. The carbon brushes should be free to move up and down. They can be removed by grasping them with the fingers and turning them clockwise as you pull them out. Clean dry fingers are in order when you remove the brushes. Solvent or grease tends to cause the brushes to deteriorate. Sockets can be cleaned with a round brass brush of the kind used to clean a rifle barrel.

Unscrew the four screws that hold the front casting in place. The coil wire will keep it from coming all the way loose. Swing the casting to one side and remove the magnetic rotor. I slide the rotor into a spare main casting to keep the magnetism from weakening. Set the rotor to the neutral point, or keep a coil in place, so that the magnetic circuit is complete. A pair of steel hose clamps, snugly wrapped around the rotor, will also function as a magnetic keeper.

A close-up view of a DRV2 magneto. I always like the appearance of this magneto; its performance these days leave something to be desired. Unfortunately, the original coil is no longer available. The replacement coil produces a spark that is inferior to the original coil. They still run all right.

Remove the Coil

Undo the nut that holds the wires to the fixed (insulated) contact. Slip the wire through the recess under the distributor gear. This wire attaches to the coil after running through a tunnel. A coil with a hard plastic cover has a stud with a nut protruding. Disconnect the wire and leave it with the main housing. If the coil is tape wrapped, or a later type coil, you will have to cut the primary lead wire if you want to remove the coil. Watch where you cut it; leave room so that you can easily splice the wire together when time comes to re-install the coil.

A snap ring holds the distributor rotor in place. I use a small, very sharp screwdriver to pry the ring out of the groove. A larger screwdriver can be used to slide it out over the cam. I follow my fathers instructions for handling snap rings or small spring clips. "First you sweep the floor. Then clean off the top of the workbench. If possible remove every other item from the room and close the door. Now this will not aid in removing the ring, but it surely is helpful when you have to look for the blasted thing," he said.

Once the snap ring is removed, slip the retainer washer off. The gear will now slide easily from its mooring, leaving the pair of number fifteen bearings behind. They always come apart easily for me. If the shaft does bind, support the housing carefully in the vise and tap the end of the shaft with a brass drift. Catch the fiber gear when it does come loose.

Remove and Test the Coil and Condenser

The condenser lives in a compartment beneath the distributor gear axle support. Screws can be removed from the top side, and the condenser is easily removed once the gear is out of the way. Test the condenser with a condenser tester. Don't be surprised if the condenser fails the test—they usually go bad. For replacement I use a 0.2 microfarad, 200 volt, axial-lead, mylar-insulated capacitor from the local electronics store. Solder a ring terminal on one lead. Be careful to "heat sink" the lead with an alligator clip or use the ever handy hemostat. Solder a wire to the other lead and insulate with spaghetti or heat shrink tube. Use super glue or epoxy to stick the body of the capacitor to the place where the original condenser was located. Solder a terminal to the other end of the wire, which is cut long enough to reach the stud where the coil wire connects to the fixed contact.

Test the coil with a commercial coil tester, or if you lack a tester, check secondary resistance with your multimeter. I don't have the specifications for the coil, which was an A2437. Check the primary for continuity and low resistance. Should be 0.3 or 0.4 ohm. If the coil was a tape-wound

Note the three keyways. This is a fits-all pawl plate. The keyway to the right is the proper one for a DVR2. When I was young and foolish, I rebuilt a DVR2 and installed the wrong impulse plate. It worked fine on my test bench where I had it mounted with the flange in the horizontal position. I finally visited the customer at his home. Fortunately, I had ordered a complete new impulse coupling. When I actually saw how it was mounted I realized there was something wrong. "It must be a defective impulse," I stammered. "We can take care of this on the spot." I fudged the new, three keyway impulse onto the magneto as quickly as I could, pretending that I knew what I was doing. It worked fine.

coil, check for softness. If it's soft and squishy, it should be discarded. New coils are very difficult to find. An R2477C or a T2477C Fairbanks-Morse coil will fit. You must bend a high tension clip so that it will touch the distributor rotor brush. You won't get quite as good spark as the original, but it should be satisfactory.

Clean the housings with a good solvent. Clean the magnetic rotor and bearings and examine the races and balls for pitting and checking. If the bearings must be replaced, be careful to reinstall the shims. Check the end play. I try to keep the range from 0.0005in preload to 0.0015in slop. Grease the bearings with good bearing grease. Fill the space in the bearings about one half full.

Reassembly

I have a special tool to hold the impulse plate when I assemble the impulse coupling. The tool has a tapered shaft with a permanently attached woodruff key. I slip the impulse plate over the shaft and woodruff key. Holding the tool in my left hand—since I'm right handed—I slide the cup

This is how most DVR2 magnetos come into my shop.
Stripped down and cleaned up in my vibrating cleaner and
glass bead blasted, it will look like new. I couldn't save the coil.

The magnetic rotor from a DVR2. The magnetic rotor was one
of the first magnetic rotors with alnico magnets cast into the
aluminum. The poles are made from laminations. Corrosion
within the laminations causes total failure.

This coil is an example of the original coil for the DRV2. The button on the top of the coil contacts the distributor rotor. An R2437C or T2437C coil can be adapted to fit. Bend the copper contact so that it contacts the brush on the distributor rotor.

with spring over the impulse plate. There are two slots in the center tube of the plate. Slide the cup down so that the bent end of the spring slides into the longer slot. Press the cup over the plate. When it hits bottom, turn it gently to the right (if the magneto was a counter-clockwise magneto, you would turn it to the left). Pull the cup gently outward, while maintaining pressure against the spring. When the lower lips of the cup clear the plate, turn the cup to the right, increasing the pressure. When the lip has cleared the extension of the plate, slide the cup down. Spring pressure should hold the cup in place. Slip the assembled impulse coupling from the tool and carefully place it on the work bench until you are ready to install it on the magneto.

"Ah ha," you say, "but I don't own a special tool."

"No problem," I reply. "Take a few minutes now and clean up the impulse coupling. As soon as you have removed the magnetic rotor, wrap a clean shop rag around it, and use the rotor as a tool to assemble the impulse. The shop towel is necessary to prevent sharp metal filings collected on the magnet from being driven into your hand.

Install the condenser, the plastic or fiber plate that fits under the distributor gear. Install the distributor gear, ignition points, wire to the condenser, and coil. Install plate. Adjust points to 0.012in. Tighten the coil in place. Attach the primary wire. Install the distributor cap and gasket. Rather than cutting a gasket, I usually put a bead of silicone sealer around the edge of the cap and allow it to dry for several hours. After I install the cap, I trim the excess sealer with a sharp knife.

Test on the test bench or the tractor.

Chapter 9

Fairbanks-Morse J4B3 (Allis-Chalmers)

The Fairbanks-Morse J-series magneto is an effective, powerful, modern magneto. It has one important weakness: The magnetic rotor tends to loosen on the shaft. Since it is important to have the points open at a precise point relative to the position of the magnet within the housing, a loose magnet defies adjustment. A loose magnet causes an erratic miss when cold and severe lack of power when hot. This condition mimics having a defective coil.

A 1939 Allis-Chalmers RC with the standard J-series magneto. The J magneto was used on many Allis-Chalmers tractors produced in the late thirties and the forties. The mostly high-quality J magneto has one weakness; the magnetic rotors loosen on the shaft. Loose magnets cause poor performance. Recently, I began to pin the loose magnets by drilling a 1/8in hole through the collar below the magnet and fasten it with a piece of 1/8in steel welding rod.

An FMJ2A39 on a standard twin. The J-series and the X-series Fairbanks-Morse magnetos use a simple code to identify magnetos. FM—surprise, surprise—stands for Fairbanks-Morse. The next letter stands for the series, in this case J series. The number indicates the number of cylinders (two, in this case).

The letter A indicates a base-mounted magneto. B stands for flange mount. The final number, thirty-nine, indicates the customer. Therefore, FMJ4B3 indicates a J series for a four cylinder engine that is flange mounted. The customer or user is Allis-Chalmers.

Always check the magnetic rotor for looseness. Grasp either end of the shaft in the vise (use soft jaws). Grasp the magnet with your gloved hand and wiggle it back and forth. If it moves at all, it must be pinned or discarded.

Disassembly

Before you can check the magnetic rotor, however, you must dismantle the magneto. Remove the two screws that hold the distributor cap on the cover. Remove the four screws that hold the cover in place. If the screw heads are rusty, be gentle. If they are very tight, warm the housing near each screw with a propane torch. The distributor rotor and its shaft and drive gear remain in the cover.

A snap ring or C clip holds the distributor drive gear to the shaft. The gear has a slot that slides over a pin protruding from the shaft. Remove the gear from the shaft. It is a good idea to slip the pin from the shaft and put it away.

Disconnect the coil, condenser, and shut-off wire from the points. Remove the points and condenser from the contact plate. Check the shaft for side play in the contact plate bushing. Take out the four screws and remove the contact plate.

Two set screws hold the coil in place. Clean out the slots and find a matching screwdriver. Use the propane torch to heat the area adjacent to the screws while maintaining pressure with the screwdriver. The coil set screws tend to stick in place and rough handling will destroy the slots. The bad

A Fairbanks-Morse J4B7A on a Wisconsin engine in a Gibson tractor. This magneto uses a two-spark coil instead of a distributor rotor. A two-spark coil fires both plugs at the same time. The seven at the end of the designation shows that Wisconsin is the user. The A designates a specific modification for a certain model Wisconsin engine, which is usually something simple like relocating the stop switch terminal to the left side of the magneto.

news is that the set screws are a pain to drill out. The good news is that the 1/4in set screw may be replaced by a 5/16in set screw if you foul up the removal of the small set screws.

Remove the shut-off wire by removing the switch from the outside of the case. Some models have a flat metal shut-off switch and others have a push button. In either case you must hold the screw head inside the housing. A knife blade or a small right angle screwdriver will work.

Impulse Coupling Removal

Turn the housing around and remove the nut holding the impulse coupling in place. Fairbanks used a washer with bendable ears to keep the nut securely in place. Flatten the ears and use a 3/4in socket and an air wrench to remove the nut. You can also wedge a screwdriver against the edge of the magnet within the housing and use a plain socket wrench. Fairbanks-Morse made a special tool to slip over the impulse cup dogs.

A two-jawed puller may be used to remove the impulse coupling in one piece, or you can remove the impulse cup by grabbing one ear with a vise grip pliers. With the pawl touching the pawl stop, turn the cup clockwise a few degrees and pull outward. When the cup clears the pawl plate, allow the spring to unwind and separate from the plate. Use the two-jawed puller to remove the pawl plate from the tapered magnet rotor shaft.

Pull the Rotor

Pick the woodruff key from the slot. Pry the seal retainer from the housing. Pry the seal from the housing using a small screwdriver. The second seal retainer will usually fall out, or can be loosened with a jet of compressed air. Behind the retainer is the hairpin clip securing the shaft to the inner race of the bearing. I use the bent end of a scribe to hook it loose. It is wise to wear safety glasses or a face shield as the clip will fly if you give it a chance.

Press the magnetic rotor from the housing. I usually wrap the rotor in a shop towel and place it where it won't pick up nuts and bolts. Check it for a loose magnet. The magnet may be pinned if the collar at the drive end is large enough. A 1/8in drill may be used to drill the hole for a 1/8in steel pin. Try to center the magnet on the shaft when

A base-mounted Fairbanks-Morse four-cylinder J series. This may or may not have come off of an Allis-Chalmers.

you drill through. If the hole isn't perfectly spaced, the E gap may be compensated by using a different points' gap. Running the magneto on a test bench and changing points' gap until the best spark is obtained is one way to compensate. Trial and error also works.

Remove the C-ring from the interior of the housing. Press the ball bearing out, clean and inspect it. The bearing is Fairbanks-Morse part #C5949. You may replace it with a Peer #7109 or equal. The Fairbanks-Morse price is rather high. I buy in quantities of ten or twenty from a bearing distributor at about three dollars each.

The stripped housing may be cleaned with solvent. Paint may be stripped in carburetor cleaner or commercial paint stripper.

J cover with the rotor, gear, and shaft in place. When you take the cover off, the gear and so forth stay with the cover. Remember to match the timing marks on the rotor gear with the drive gear on the end of the magnetic rotor.

Cleaning and Testing

Examine the coil carefully for carbon tracking. If it is a tape-wrapped coil, check for leakage or softness. If you have a commercial coil tester, test for leakage, resistance of primary and secondary windings, and spark. The coil is Fairbanks-Morse #R2477C. Replacement coils in both original and aftermarket manufacture are available from Standard Magneto Sales Company in Chicago.

Return to the distributor cap. Remove the brush (part #E2460B) and put it safely away. Examine the cap for carbon tracking and wear on the points. Clean the sockets where the high tension wires plug in. Use a wire brush or a strip of sandpaper fastened in a split in a dowel. Perfec-

Inside the J cover. The timing marks are not visible. I find that a drop of solvent on the gear may make the timing marks visible. If you still can't see the marks, you can transfer the timing marks from a good gear. Should worse come to worse, set the rotor so that it is pointing to the number one cylinder. Turn the magnetic rotor to a neutral point and slip the cover in place. I used to recommend replacing the fiber timing gear and shaft to be on the safe side. Now they want over $40 for the part. If it is intact, use it. I used to say, "When in doubt, throw it out!" No more.

tion is not required at the high tension end of the magneto. Perfection is required when cleaning the points. Voltage at the high voltage end of the system is high enough to overcome a little corrosion. At the low voltage end of the system even a couple of ohms is too much.

If the high-tension points are severely worn, the cap (part #C800) must be replaced. If the distributor rotor (part #M2765) has a badly eroded high-voltage conductor, or is cracked, it should also be replaced.

The cover (part #CX2430) contains the bearing for the distributor gear and shaft. Over time, the porous bronze bushing loosens in the cap al-

lowing the gear to be pushed out of position. The easiest repair is to replace the cover which is a twenty-five dollar part (the last time I looked). The bushing (part #B5950C) is available. Getting it to stay in position is the trick. There is a felt washer in a recess beneath the distributor rotor. I always put a few drops of oil on this felt. Although the bushing is separated from the felt by a steel washer, a dry felt tends to wick oil from porous bronze bushings.

The original fiber gears had a plastic washer fastened to the outside surface with instructions and indicator marks about timing the distributor gear to the timing gear. If this plastic is missing

The driven end of the J magneto. You can tell the number of times the magneto has been overhauled by noting the different tool marks where the seal retainer is staked in place. I see two sets of marks on this housing. The original factory marks are rectangular. A round punch was used on the overhaul.

A special tool for holding impulse dogs. This tool has been around for many years. I think that Fairbanks-Morse was the supplier. There is no name on the tool. Wico (now Standard Magneto) supplies a similar device. It come in handy from time to time.

An impulse from a J-series magneto. The pivots for the pawls were riveted into place. Sometimes they become loose. Peening them into place doesn't work for me. I usually replace them with good used parts or a new replacement impulse. If a pawl becomes loose when the engine is operating, it may fly right through the side of the magneto case. Operating an engine for an extended time at a speed where the impulse is still clicking usually causes this catastrophic failure.

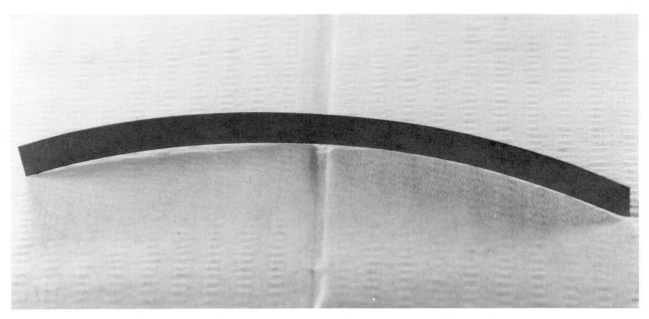

Part #A2824, a bearing shim. The main (C5949) bearing in all the early J series used paper shims. The engineers were trying to prevent stray electrical currents from going through the ball bearing. Electricity can cause the ball to develop pits and a short life. Magneto bearings were commonly insulated dur-ing the twenties and thirties. Some Lucas motorcycle magnetos used the shims until about 1960. Everyone else, as far as I know, dropped the practice during the late forties. These shims are a real pain to install properly.

you will have to compare the relationship of the flat on the shaft to the marked teeth and transfer the mark to the old gear. A *C* or *A* is marked directly on the fiber gear on later model and replacement gears. A drop of solvent on the gear face, a bright light, and a magnifying glass will usually make the marks visible. A sharp scribe will make a permanent mark.

Reassembly

Install the distributor rotor on the distributor shaft before assembling the cap to the magneto. Support the gear with your thumb when you press the rotor on the shaft. When the time comes to install the cover to the housing you must align the mark—in this case, the tooth marked with a *C*—so that it will match the tooth marked at its base with a dot on the drive gear. You slip it into place while hoping that nothing turned in the process. If you can't see the mark on the drive tooth, turn the magnetic rotor to the neutral point. Aim the driven gear mark so the *C* tooth will contact the drive gear.

Pack, or replace the driven end bearing. Install the snap ring. If the off switch has been removed, clean and install the wire, insulator, terminal screw, insulation bushing, insulation washer, off switch, and lock washer and nut. Press the magnetic rotor into place. Install hairpin retaining clip, inner seal retainer, oil seal (part #G3861), and

the outer retainer. Stake outer retainer in place. Since I like to re-magnetize the rotor in the housing, I usually leave the impulse for last.

Assembling the Impulse Plate

I have a special tool to hold the impulse plate when I assemble the impulse coupling. The tool has a tapered shaft with a permanently attached woodruff key. I slip the impulse plate over the shaft and woodruff key. Holding the tool in my left hand—since I'm right handed—I slide the cup with spring over the impulse plate. There are two slots in the center tube of the plate. Slide the cup down so that the bent end of the spring slides into the longer slot. Press the cup over the plate. When it hits bottom, turn it gently to the right (if the magneto was a counter-clockwise magneto, you would turn it to the left). Pull the cup gently outward, while maintaining pressure against the spring. When the lower lips of the cup clear the plate, turn the cup to the right, increasing the pressure. When the lip has cleared the extension of the plate, slide the cup down. Spring pressure should hold the cup in place. Slip the assembled impulse coupling from the tool and carefully place it on the work bench until you are ready to install it on the magneto.

"Ah ha," you say, "but I don't own a special tool."

"No problem," I reply. "Take a few minutes

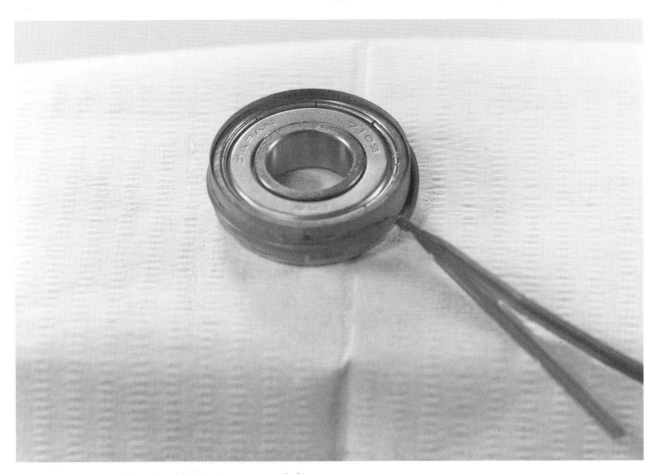

An easier way to install bearing shims is to store a couple fastened around the outside of an old C5949 bearing with a rubber band or tag wire. A thin coat of grease helps hold them in place.

now and clean up the impulse coupling. As soon as you have removed the magnetic rotor, wrap a clean shop rag around it, and use the rotor as a tool to assemble the impulse. The shop towel is necessary to prevent sharp metal filings collected on the magnet from being driven into your hand.

Install the coil, contact plate, points, condenser, and distributor drive gear with its retaining clip. If you removed the pin, be sure to install it before you slip the gear on the shaft. Adjust the points to 0.020in. If you have access to a magnetizer, re-magnetize the rotor. Install the impulse

coupling and woodruff key. Bend a tab up around the nut.

I like to test the magneto with a grounded screwdriver at this time. If it snaps a 1/4in (or larger) spark using the impulse coupling, I feel confident that the magneto will perform on the tractor. I run it on the test bench anyway. I like to see it coast down all the way without dropping a spark. If it misses a few sparks just before the impulse catches, I take it back to the drawing board, or workbench and find out why.

Chapter 10

Fairbanks-Morse X4B9

The X-series Fairbanks-Morse magneto is the successor to the J-series magneto. Fairbanks-Morse remedied the weaknesses of the J-series magneto in the attachment of the magnet to the rotor shaft, and by relocating the distributor shaft to a bushing in a boss attached to the contact support plate. Unlike the J series, the distributor timing gear can be assembled accurately to the rotor drive gear before the cover is installed.

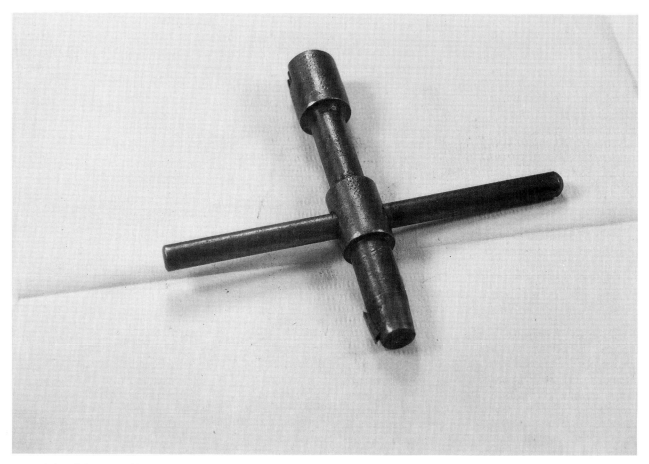

A special tool for assembling Fairbanks-Morse impulse couplings. One end has the proper taper and a keyway for impulse couplings, the other end has a fitting which looks like it is used for tightening springs. This is a very handy tool for in-stalling the cup to the pawl plate. I don't know who made this tool, and I suspect that they are very hard to come by. You can make a good substitute tool by demagnetizing an old magnetic rotor and gluing a woodruff key in place.

On the downside, now you occasionally find a distributor rotor that is short circuited from the center contact to the rotor shaft (if you have a good spark with the cover removed, but none with the distributor cap and rotor in place, try a substitute rotor). I test for this condition with my high voltage tester.

Disassembly

As always, I remove the fragile parts first. Two screws release the distributor cap (same part as used in the J series). Remove the distributor rotor from the shaft by gently prying with small screwdrivers (rotor M2765 is the same part number as used in the J series). Sometimes corrosion holds the rotor tightly to the shaft. The main cover can be lifted straight out: The hole in the center of the cover is too small to let the rotor remain in place, so the rotor comes with the cover, or the cover breaks. Be gentle! The boss that supports the distributor rotor shaft is not very strong. Sometimes you can cut part of the rotor away, and break the rest off with a screwdriver or pliers. I always put a drop of lithium grease, or anti-seize compound, on the top of the rotor shaft when I install the rotor.

A special tool for removing and installing the C5949 ball bearing in J- and X-series Fairbanks-Morse magnetos. The top end of the tool will fit through the oil seal recess and push on the bearing. To install the new or cleaned bearing, insert the bearing over the lower end of the tool. The outer edge of the tool exactly matches the race and the length of the tool is just right for pressing the bearing into its recess.

A special tool for removing and installing castle nuts on the driven end of Fairbanks-Morse magnetos with extended magnetic rotor shafts used on Wisconsin and Kohler engines. Spe- *cial tool kits used to be available from the various manufacturers of magnetos. Alas, this is no longer the case. Flea markets are a possible source of these old and very special tools.*

The X-series magneto is quite similar to the old J-series. The magnet is fastened more securely to the shaft and the distributor drive gear eliminated the old tiny locator pin in favor of a flat spot on the shaft. They added a boss to the contact plate with a bushing to hold the distributor gear which now remains attached to the magneto body when you remove the cover. The contacts were redesigned. The impulse pawl plate pivots are more securely attached to the pawl plate. With the cover and cap installed, the J- and X-series appear identical.

Remove the points and the condenser from the contact plate. Extract the snap ring from the distributor gear shaft and pull the gear from the shaft. Remove the contact support plate from the housing. Slip the coil from the housing. The coil retaining screws are 1/8in, which remove easier than the 1/4in size used in the J series. It still pays to heat the housing next to the screw with the propane torch or a heat gun before removing them. I always exert force with a screwdriver as I heat the housing. The screw loosens magically.

Pull the Pawl Plate

Use a small screwdriver to bend the retainer washer lugs away from the nut. In 1970, Fairbanks-Morse began to use a shorter, self-locking nut with a plain washer to replace the original nuts. The original nut was part #K2570, the washer A5931B. The new nut is U2570 and the washer B5931B. I usually reuse the old nut. If the lock washer is in bad condition, I use Loctite. Remove the nut. Turn the impulse until one of the pawls touches the stop. Grasp the impulse cup and turn it slightly to the right, and pull out until the cup lips clear the pawl plate. Unwind the spring and remove the cup. With a two-jawed gear puller, remove the pawl plate. If the pawl plate is stuck, try heating it with the propane torch. Tighten the puller until the plate is under tension. Sharply strike the puller screw with a medium hammer. Usually the pawl plate will pop loose.

After removing the pawl plate, remove the woodruff key. Pry the seal retainer from the housing. Remove the seal, inner seal retainer, and hairpin clip. Press the magnetic rotor from the housing. Turn the housing over and remove the ball bearing retainer snap ring. Press the ball bearing from the housing.

With the Dremel Moto-tool, or a utility knife, remove the staked metal protruding from housing where the outer seal retainer was staked in place. If you are going to dunk the housing in a strong cleaning solution, remove the shut-off switch terminal screw, nuts, washers, and insulators.

Testing and Cleaning

Check the magnetic rotor for a loose magnet (never take anything for granted). Test the coil and condenser. Clean the points, if you intend to reuse them. Clean all the parts in solvent, rinse in hot water, and air dry them. Clean the ball bearing. The X-series magneto uses the same bearing (C5949 or Peer 7109) as the J series.

Inspect the impulse coupling for excessively-worn pawl pivots. At this time, I like to demagnetize the cup and the pawl plate by placing them one at a time in the V block of the growler. Turn on the growler and remove the parts slowly from the magnetic field and turn off the growler. Do not run the growler for more than five or ten seconds at a time. Check for remaining magnetism by moving the parts near a compass and turning it while watching for deflection. Assemble the impulse coupling. You may use the magnetic rotor shaft as an aid. Be careful not to touch (or even get close) to the magnet with the impulse coupling parts.

I use a dummy shaft tool for this assembly. You can use an old magneto shaft for this purpose by demagnetizing the magnet with the growler.

You may have to demagnetize several times to kill the magnetism. Alnico magnets are very powerful.

Check the contact plate bushing for wear. When in doubt, replace the bushing (part #A5950A). Simultaneously, check the distributor shaft bushing (part #D5950C) for wear. I like to load the porous bronze bushings with oil by balancing the bushing on my thumb and filling the

Fairbanks-Morse X magneto with the impulse removed.

Fairbanks-Morse X magneto with the impulse and inner seal
removed.

Fairbanks-Morse X magneto with the impulse, seal retainer, seal, inner retainer, and key removed.

Three floating drive disks. All base-mounted and many flange-mounted magnetos have a floating disk between the impulse coupling and the engine drive member. The floating disks allow slight misalignments between the magneto and engine. Steel drive disks have been used in the past but currently fiber or phenolic disks are used. They come in different thicknesses and sizes to fit various machines.

interior of the bushing with SAE 20 oil. Pressure with the index finger, or the thumb from the other hand, will force oil through the pores. When I see oil oozing from the side of the bushing, I know that there is adequate lubrication.

Clean the rust from the coil armature and from the part of the magneto where the coil is mounted. Install the ball bearing and the snap ring. If possible, magnetize the magnetic rotor and immediately install it. Put the distributor gear and shaft in the contact support plate and install the snap ring. Install the coil and the contact support plate in the housing. Install the shut-off switch terminal with its insulation washers and spacers.

Install the hairpin clip, seal, and retainers. Install the woodruff key and the impulse coupling. Install the washer and nut assembly and tighten. Bend the ears of the lock washer to secure the nut.

Install points and condenser. Adjust the contact clearance to 0.015in. Turn the distributor gear so that the C tooth is next to the rotor shaft. Install the gear so that the beveled tooth fits into the tooth marked with a C. Install the snap ring.

At this point, I like to chuck the magneto in the vise and use a screwdriver to check the spark. If it produces a 3/8in spark on impulse, there is a good chance that it will work in operation.

Install the cover gasket and the cover. Install the felt washer, distributor rotor, distributor cap gasket, and the distributor cap. Test on the test bench, if available, or install on the tractor and test.

Chapter 11

Wico, Series C (John Deere)

The Wico Series C magneto is a good, but not great, magneto. It uses the most elegant lubrication system of any magneto I have repaired. Most parts are available for this magneto through John Deere Dealers, Prestolite-Wico Dealers, and through the manufacturer, Standard Magneto Sales Co. of Chicago, Illinois.

The first step in the restoration of a Wico C is to wipe off the exterior. Before dismantling the magneto, I like to turn the magnetic rotor by hand so that I can determine the condition of the rotor and bearing. Feel for side play and the amount of magnetism. Try to find the neutral point in rotation. If everything feels normal, that is, the rotor doesn't bind or slop from side to side, and there is a definite feeling of magnetic drag through the neutral point, I will usually install it on the test bench. I remove the distributor cap before the test so that I can watch the operation of the ignition points. If the magneto produces spark, I reinstall the distributor rotor and cap and try again. If everything at this time is functioning normally I call the customer and ask for more information. Should the magneto have a history of failing only when hot, I want to know that fact.

Most of the time no spark is produced. If this is the case, I will insert a screwdriver and short circuit the moveable contact while the magneto is running on the test bench. If this causes the coil to produce a spark, then I know that the points are dirty or tarnished. On the other hand, if I get no fire at all, then it is time to dismantle and test each component separately.

If you haven't a test bench handy, you can still fasten the magneto in the vise. Carefully clamp the lower mounting ear. Always be careful how you grip a magneto in a vise. The aluminum housing is fragile. It is easy to bend and hard to straighten. You may check for spark by installing a short spark plug wire in the coil tower and hold the other end near the vise. All John Deere magnetos turn counterclockwise. Wear a glove to turn

the impulse in that direction. When it snaps you should get a spark. If the spark can jump 3/16in (or 5mm), the engine should start. Of course, you are going to tear the magneto down anyway, so if it doesn't spark, it is no big deal. If it sparks before you work on it, but doesn't when you are done, then you are in big trouble!

A base-mounted Wico C magneto used on early John Deere tractors.

Disassembly

Remove the distributor cap, gasket, and rotor and place them in a safe place. Place the magneto on the bench where you can examine it more thoroughly. At this time, take your multimeter and set the dial for resistance times one ohm (Rx1). If you have a digital meter, set it to the lowest resistance setting. This is usually the two hundred ohm scale. Zero the analog meter scale as described in chapter three. With the digital meter, note the resistance indicated when the leads are touched together. Measure resistance from the contact support plate and the main body of the magneto. The resistance should be zero. If resistance is about 0.3 ohm, turn the rotor until the points are open. If the resistance increases substantially, it means that the contact support plate has lost its ground connection.

Remove the screw that connects the condenser, moveable contact, and primary coil wire together. Remove the four screws that hold the main cover to the housing. Be careful to allow the coil primary wire to slip back with the housing.

Front view of a C cover. Note stamped type indicating point gap setting.

Turn the cover over and inspect it for signs of carbon tracking from the coil contact toward one of the mounting screws. Observe the pointed copper contacts riveted to the contact support plate. These contacts touch two of the screws that hold the laminations to the housing, completing the ground circuit of the primary winding. Frequently, corrosion forms at the joint between the aluminum contact plate, the steel rivet, and the copper strips. I have never been able to re-establish this connection. Not to worry. If the cover is otherwise in good condition, you can install a ground wire between the contact support plate and the housing. I usually run one from one of the condenser hold down screws to one of the coil retainer screws.

Remove the two screws that hold the coil retaining brackets. Turn the magnetic rotor to the neutral position. The coil will come right out. If you have a coil tester, now is a good time to test the coil. If you lack a coil tester, you can check for continuity and resistance with your multimeter. The primary winding should have about 0.3 ohm resistance. The secondary winding will be around 7000 ohms. If the coil feels spongy to the touch, or if there are signs of black gunk leaking out of the coil, the coil should be replaced.

Condenser Testing

If you have a condenser tester, test for open, short, capacitance, and series resistance. Some of the new digital meters measure capacitance. An analog meter set to Rx10,000 ohms (Rxl0k) will give an indication of condenser condition. When you touch one lead to ground and the other to the condenser lead, the needle should move from the infinite resistance position and then gradually return to infinite resistance. If it doesn't return to infinite or if it never moves at all, the condenser will have to be replaced.

Impulse Removal

Place the main cover with the cap and rotor in a safe place. Now we can start at the other end. A 3/4in thin wall socket will fit the impulse nut and a quick burst using the air wrench will loosen it in a trice. If you haven't got an air wrench, don't worry. Since this magneto rotates in a counterclockwise direction, slowly turn it until the impulse pawl locks against the pawl stop. The nut has normal right-hand threads and rarely sticks, unless someone has been abusing it with a hammer.

Remove the impulse cup by grasping a lug with a vice grip pliers. I usually just use my bare hand, but be warned, the cup has spring pressure against it and if it slips, it can give you a nasty scratch. Anyway, grasp the cup and, with the impulse pawl tight against the stop, turn it a little way to the left and pull it away from the magneto. Turn it slowly to the right, releasing the pressure of the impulse spring. When the pressure is relieved, the cup and its spring will come loose from the notches in the washers at the shaft.

Remove the washers from the shaft and then remove the plate that covers the pawls and pawl plate. Note the small washer on top of this plate. Remove the spacer between the pawls. Use a pair of screwdrivers to wiggle the pawl plate along the shaft until you can grab it and slide it all the way off of the shaft. Remove the four screws and star washers that hold the pawl stop plate in place. Remove the pawl stop plate and gasket. At this point, you can easily push the magnetic rotor out through the other side of the magneto.

If you don't have easy access to a magnetizer, fabricate some sort of a "keeper" to prevent the magnetic rotor from losing magnetism. A pair of steel radiator hose clamps, fastened around the rotor will probably work. Place the magnetic rotor at a distance from nuts and screws. Otherwise you will have to keep pulling metal parts from the magnet. Be careful when you handle any magnet since they tend to pick up tiny slivers of steel and bury them in your hand.

Removing the Oiling System

Remove the oil disk. It should easily slide off the end of the shaft. Pull out the oil scraper and spring for the recess. Remove the felt crescent from the oil reservoir. Use a tweezers or small needle nose pliers to remove the copper device that presses the oil retaining felt against the oil disk.

In operation, the oil disk rotates past the oil-soaked felt, picking up a film of oil. The oil is scraped from the disk by the fiber scraper and flows through the passage ending between the two bushings that support the rotating magnet. Oil that passes toward the driven end runs into the oil disk, where it is picked up and circulated through the oil passage and bushings. Oil that passes through the other bushing is picked up by the magnetic rotor, where it is moved by centrifugal force to the edge of the rotor. When the oil is thrown from the rotor it is captured by a ring. Gravity pulls the oil downward where it flows through a square hole into the oil reservoir and into the felt. Capillary action will eventually return the oil to the oil disk, and the cycle repeats itself.

Two things are noteworthy here. The reservoir with its felt insert holds a lot of oil. The oil is circulated through the main moving part of the magneto and after use is recaptured and used again.

There is a handy plug screw on each side of the magneto body. If you use a pistol oiler to

Backside view of the C cover. The pointed copper contact riveted to the contact plate. The other contact is not visible. These two contacts connect the aluminum contact plate to the main housing. The steel rivet will sometimes corrode and the connection between the contact plate and main housing is lost. A wire may be run from the condenser ground to the coil retaining bracket to complete the circuit and save the cost of replacing the cover.

squirt oil in the reservoir, feel free to squirt all you want to. If you attempt to overfill the reservoir, the excess oil will drain through a handy drain hole in the bottom of the casting and do no harm. In the past twenty years I have repaired a goodly number of the C Wico magnetos. It is rare to find one that has completely dried out. When you find one with a bone dry felt, you usually find worn bushings that need to be replaced.

Remove the Bushings

To remove the bushings, use a long tapered punch inserted through the center of one bushing to tap the edge of the other bushing. Very little force is required. After the first bushing is removed, insert a straight punch which will fit through the tube but not the bushing at the other

A flange-mounted C case. The coil has been removed. The rotor is in place and set in the neutral position. One pole of the magnetic rotor is at the top—the other pole is at the bottom.

Magnetism is short circuited from top to bottom on each side. If the coil had been in place, no magnetism would go through the coil.

end. Again, a light tap is all that is required. Be careful in handling the housing so that you do not damage the tube.

Never remove the laminations from the housing—it is a real pain to reinstall them. Nothing is gained by removing them.

The porous bronze bushings are hat-shaped with a small fillet between the crown and the brim. There is a notch in the tube to match the fillet. When the bushing is properly installed, the notch holds the fillet and prevents the bushing from turning. The hat brim handles thrust.

Cleaning Parts

Stoddard or some other type solvent and a soft parts brush will soon clean the grunge from the parts. Rinse in hot water and air dry. Soak the felt in solvent or alcohol and dry with a paper towel. I usually rinse the felt several times until I have squeezed out all of the dirty old oil. When it is satisfactorily clean, submerge the felt in a container of SAE 20 oil or whatever similar lubricant is handy. Three-in-One light motor lubricant has worked well in the past for me. I suspect that any modern engine oil would work as well.

If you have access to a magnetizer, magnetize the magnetic rotor and immediately install in the housing. Leave it in the neutral position until you have tested and or replaced the coil. Install the copper spring device and the well-oiled felt. Install the scrapper in its recess. Install the oil disk. Oil the leather seal in center of the pawl stop. The metal bushing usually sticks with the seal. Do not try to remove the seal. These seals are hard to find. Fortunately, they wear like iron.

Install the pawl stop plate (be sure to have the pawl stop to the left side). Note the index mark at the top of the plate. Observe also the series of marks on the magneto body directly above the pawl stop plate. These marks are used with the witness mark at the top of the pawl stop to set the proper lag angle of the impulse.

Set the Lag Angle

The top center mark indicates a lag angle of fourteen degrees. Each mark is five degrees apart. This magneto requires a lag angle setting of thirty degrees. Move the witness mark to the left by three marks (fifteen degrees past top center equals twenty-nine degrees) plus a tad. The tad should get us to thirty degrees. Tighten the four mounting screws.

Install the pawl plate on the shaft with the pawls facing you. Place the spacer over the shaft, between the pawls. Install cover plate over the pawl plate so that the pawl pivot shafts are in the two holes. If they don't fit, remove the cover plate and turn it over and reinstall. Install the washer on top of the cover plate. Place the two notched heavy washers on top of the washer.

The Impulse Coupling

Examine the impulse coupling cup and spring. If the spring is broken or badly rusted, pry it out of the cup. Note the direction of rotation of the spring. Install the new spring in the cup. If the old spring is in good condition, continue to the next step.

Place the magneto on top of the vise, with the drive end up. The jaws of the vise should be open far enough to allow the laminations to set down between them but not allow the magneto to fall

The outer end of the bushing is visible. The light-colored ring at the end of the bearing tube is actually the brim of a hat bushing. A hat bushing looks like a stovepipe hat; the top and bottom of the hat is open. The brim of the bushing handles end thrust. The notch in the brim fits into a matching notch in the tube. This keeps the bushing from turning in the tube. The dark opening below the tube allows oil to drain back to the oil wick. The threaded hole in the side of the case—halfway between the flange and the cover—provides access to the oil wick.

The felt wick in the upper left corner holds at least a table-spoon of oil. The horseshoe-shaped piece below and to the left is a spring copper device that presses the oil wick against the thin disk. Oil carried by the disk to a passage above the center bushing is picked off by the spring-loaded wiper located to the left of the disk. The oil travels down a passage to the area between the inner and outer bushing. Oil leaking past the bush-

ing at the driven end is directly captured by the felt. The pawl stop and oil seal is below the spring and impulse cup in the picture. The housing in this case is BC (before cleaning). Visible through the lower mounting hole is a final drain hole. If you go wild with your pistol oiler and overfill the case with oil, excess oil runs out of the drain hole.

through. Grip the laminations with the vise ever so gently or just tight enough to keep the magneto from sliding around. Now for some fun!

Turn the impulse plate to the left and allow one of the pawl to engage the stop. The trick now is to engage the curl in the end of the spring with the notch in the two washers, and to place about a half turn of spring pressure counter-clockwise by turning the cup. The cup should now be pressed down in position. The edge of the large notch in the cup should engage the large ear of the pawl plate. Install the nut, preferably with the air wrench. When you try to tighten the nut, the magnetic rotor will turn. The air wrench comes in handy here. If you don't have an air wrench, hold the other end of the shaft at the D-shaped part, either with the vise (soft jaws, if possible) or with a set of vise grips.

The first time I installed an impulse cup on a Wico magneto, it took an hour or two. Now I usually slip them right into place. Every so often,

an impulse fights back and I have to wonder if I ever learned anything. One trick I use is to set the magneto in place with the notch of the washers facing me. I slip a small screwdriver behind one turn of the spring and pry the end loop down a little. I squint through the center of the cup and line the loop up with the notch and force it down until it is firmly in the notch. Then I turn the cup to the left to put a little tension on the loop. I slowly extract the screwdriver and press the cup into place, firmly seating the loop all the way to the bottom of the notch. Turn the cup until the edge of the cup is ready to disengage the pawl. Pull the cup away from the magneto until you clear the pawl with the edge of the cup. Turn the cup until it clears the ear and can be pushed down in place. Install the nut. When you twist the coupling, the pawl should engage and tension should increase by a fair amount. When the cup trips the pawl, the magnetic rotor should snap around briskly.

Install Coil, Condenser, and Points

At the other end, install the coil. If a ground wire is needed, install one end now. Install the cover gasket and the cover. Fish the coil primary wire over the top of the contact support plate. Install the points and condenser and fasten the wires to the insulated pad that is part of the condenser bracket. Loosen the contact hold down screw and turn the eccentric adjustment screw to adjust the opening when the cam lobe is directly under the contact rubbing block. The opening should be 0.015in.

A reminder—when you install a set of points in a magneto, clean them carefully with alcohol, lacquer thinner, or electronic contact cleaner. There is a preservative on the points to prevent tarnishing. This must be removed. If the points appear to be dull colored, they are probably tarnished. Polish them with jeweler's rouge and wash off with alcohol, thinner, or contact cleaner. If the points are new, you should be able to see your reflection in them. POINTS CANNOT BE TOO CLEAN!

Test Your Work

Test the spark on impulse by inserting a wire into the cover and holding it 3/16in from the housing. If you get good spark at this point, install the distributor rotor and cap with gasket. If you have a test bench, test it. If you have a tractor, install it and try it out.

If you are not going to be able to put the magneto into service for a protracted time, put a drop of oil on the points. Remember to carefully clean them before placing the magneto in service.

In case the spark is not of the quality desired, remove the cap and rotor. If you hold the magneto upside down the pawls will not engage. Turn the impulse cup until you feel the neutral point. Check the points. They should be closed. Turn the impulse counterclockwise approximately ten degrees. The points should just open at this point. Disconnect the coil wire from the insulated bracket. Set your multimeter to Rx1 ohm. Attach one lead to the moveable contact, the other to the case. Set the magnetic rotor to the neutral point. The points should be closed and the meter should read zero resistance. When you turn the rotor in the direction of rotation, look past the contact plate at the magnet. When the magnet passes the neutral point by about ten degrees you will see the edge of the magnet with a clearance of approximately 1/8in from the laminations. The points should just start to open. If they are still closed or if they had opened before the ten degree point, adjust the points until you achieve the opening at the proper point.

If you have to adjust the points to achieve this point of opening, measure the amount the points are open when the lobe is at its highest point. If the gap falls between 0.010 and 0.025in you can get by. If the gap is greater or smaller that this amount, it is time to go back to the old drawing board. The contact rubbing block may be worn. If it doesn't compare with the rubbing block of a new set, the points will have to be replaced. If not, examine the cam carefully, it may be too worn to function properly and the magnetic rotor may need to be replaced.

With luck, the magneto will now function perfectly for a long, long time. Remember to oil it once or twice a year, depending upon use. If you run the engine for a good long time, the points will become broken in, and show a nice gray spot where they come together. Well broken-in points give fewer problems than new, but unused, points. If the tractor is hard to start after sitting for a long time, take a few minutes and clean the points.

Another view of the housing. On the left side of the oil reservoir you can see the opening to the oil access holes. The white spot below the center magnetic rotor bearing boss is the opening to the inner housing oil capture port.

Chapter 12

Case 4JMA Magneto

One of the sadder days of my life was the day that I ordered a whole bunch of Case Magneto parts and was told that Slick Manufacturing had discontinued making parts for the old Case magnetos. Particularly, the plastic parts such as coil covers and distributor caps. In the magneto repair business, as in most repair businesses, there are parts known as "fast-moving" parts. These are the parts that are fragile or wear out. These parts are money-makers. When they are no longer

A Case tractor at work. Flat belts have always fascinated me. I don't understand how they keep the belts on the pulleys when I have trouble keeping V belts in place.

Case tractor with a Case 4JMA magneto. Case Magneto company also made aircraft magnetos. When I became an FAA-certified repair station, the inspectors examined my large supply of special tools for Bendix Magnetos. "Lets see the special tools for Slick magnetos," he said (Slick Manufacturing had purchased Case Magneto Company). I fumbled in my pockets and finally pulled out a nail, an adjustable (crescent) wrench,

and a hammer. The inspector looked carefully at the nail, and then he rubbed his chin. "Looks good to me," he said, "lets look at the Eisemann tools next." The nail fits through the case and the magnetic rotor shaft for the purpose of setting the E gap. Later of course, I showed the inspector a whole box of special tools for Case and Slick magnetos. The key tools were the special coil installation and removal tools.

available, for practical purposes, the device is dead. I stocked up on Fairbanks-Morse model X4B9A magnetos, a direct substitute.

Disassembly

If you have a Case magneto that appears to be in fair condition, an attempt to repair it is in order. The 4JMA magneto has some interesting features. The impulse stop is a plate similar to the Wico system. In this case (no pun intended), remove the pawl nut and the impulse nut from the rotor shaft. Pull the impulse using a two-jawed puller. Remove the snap ring and note the lag an-

gle markings in the housing. The lag angle can be adjusted by removing the pawl stop and replacing it so that the witness mark points to the desired number of degrees of lag.

Before you dismantle the impulse coupling, first remove the distributor cap and then the coil cover. Put the covers in a safe place. Loosen the small screw in the distributor gear support plate and slide the distributor rotor out. Note the two beveled teeth next to the distributor plate. When you install the plate, be sure to match the punch mark in the drive gear with the beveled teeth of the rotor.

Cam and Rubbing Block

Remove the condenser and the points. Slip the snap ring from the magnetic rotor and pull the gear. My battery lug puller does a good job in removing the gear. Remove the washer and the cam. Be very careful when removing the cam since it is made of fiber. In the old days, a fresh cam was included with the points. Note that the rubbing block on the moveable contact is steel. By using a steel rubbing block on a fiber cam, instead of fiber rubbing block on steel cam, Case eliminated cam wear as a long time problem—at least as long as new cams were available. High-time magnetos often show excessive wear on the cam as well as the surface where the oil seal rubs. Magnetic rotor replacement is the only reasonable solution in these cases.

Remove the remaining screws in breaker plate assembly and remove the breaker plate. Make sure that the woodruff keys on both ends of the rotor have been removed and press the magnetic rotor from the housing. The early 4JMA magnetos used New Departure 77502 double-shielded ball bearings. These were eventually replaced with standard three-piece magneto ball bearings. If you replace the shielded bearings with magneto bearings, you will have to shim the new bearings until you have 0.001 to 0.005in end play.

Pull the Coil

Removing and installing coils in the Case magneto is the most fun of all. Case interleaved the coil core laminations with the magneto case laminations (see sketch). In order to safely remove the coil, you need a puller as shown in the diagram. The arbor part rests on the magneto case.

Pivot the puller hooks under the coil laminations, next to the coil. Tighten the screw and the coil will come out. If you don't use a puller, prying the coil out is a real chore. Pressing the coil back into place is almost as much fun.

Reassembly

Now that you have replaced the coil as well as cleaned and lubed or replaced the bearings, assemble the magnetic rotor to the case. Tap the woodruff keys into place and slip the fiber cam over the end of the shaft and carefully press it into place. Put the two spacers on the shaft and install the gear and snap ring. Lubricate the bronze bushing with a few drops of oil. Install the points and condenser. The condenser should test at 0.4 microfarad. Set the points at 0.015in.

Wipe the distributor rotor with a soft rag. If there are black deposits on the rotor disk, try cleaning it with a soft eraser. Clean with alcohol or lacquer thinner before installation. Match the beveled rotor gear teeth to the dot on the drive gear. Tighten the screw that holds the rotor in place.

If you have access to a magnetizer, magnetize the rotor in the case with the coil in place. Install the woodruff key and the impulse coupling. Test for spark by installing the magneto in a vise and snap the impulse while holding a grounded screwdriver about 3/8in from the spring at the end of high tension lead.

Clean the inside of the cap with a soft cloth. If it is very dirty use a rag with solvent after removing the brushes. Install the brushes. Install the cap and the cover. Test on a test bench or on the tractor.

Chapter 13

MRF4A322 American Bosch (Minneapolis Moline)

One day a scruffy gent stopped in my shop. He was carrying a couple of heavy boxes in his arms. "Would you be interested in buying a pair of magnetos? New old stock. They fit Minneapolis-Moline tractors."

"How much?"

"Make an offer," he replied.

I looked the boxes over, and opened one. They had never even been opened, that was sure. "Twenty-five bucks a piece," I said.

"Sold."

I opened the till and gave him the fifty dollars. "How did you come by these?"

"Some guy from White Farm Equipment was throwing them out," he said. "Actually, he had six or seven. I was going to grab them all, but they were pretty heavy, and I didn't know if they were worth anything."

"I'll take any new magnetos you run across," I said, as he left.

Well, he hasn't come back, but who knows? The thing I never understand is why companies throw really good stuff away, just because it hasn't been in production for ten years. I can understand why companies like to get rid of slow-moving inventory, but why throw it away when I—and others—would be willing to pay good money for it. Most of the tractors that used these magnetos are now classics.

My old company manufactured wiring harnesses for Minneapolis-Moline in the sixties, and I enjoyed delivering them to their Lake Street Plant in Minneapolis. The company fell upon hard times and was acquired by White Farm Equipment Company. The Lake Street Plant was closed down, and many good workers hit the street. Their pensions were not adequately funded so that the workers not only lost their jobs, but their pensions were paid out at twenty-five or thirty cents on the dollar. There is a shopping mall where the plant used to stand and I feel sad every time I drive by.

American Bosch became a part of United Technologies and, a few years ago, United Technologies made an agreement with Bosch of Germany, to stop using the "Bosch" name. Bosch manufactured magnetos in New York City for many years before World War One. During the war, I have been told, anti-German feelings were so high that the company had to put "American" in front of the "Bosch" in order to survive. Then, the New York company was split from the parent.

American Bosch built very fine magnetos indeed. With any reasonable care most will run forever. The only disadvantage to American Bosch magnetos is the high cost of replacement parts.

Disassembly

The MRF4A322 is a large clockwise rotation, standard flange magneto that fires four cylinders. I usually start by removing the contact plate cover and the six screws that hold the main cover in place. Make sure that the distributor brushes are in place. Put the cover in a safe place. Disconnect the coil and condenser wire from the contact plate. Remove the two screws that hold the plate in place and remove the plate. The screw in the center of the cam comes out next. Slip the cam from the shaft. Note that there are two notches in the base of the cam. One is marked A the other is marked C. Depending upon the direction of rotation, you place the proper notch over the end of the key. That is, if the magneto spins clockwise, install the C slot over the woodruff key. If the magneto turns counter-clockwise, install the A slot over the woodruff key.

Remove the disc that covers the metal drive gear when you slide the distributor rotor and gear out. Use a small two-jawed puller to pull the metal gear from the shaft. Tap the woodruff key from its groove.

Remove the Impulse Coupling

I usually use a heavy, common screwdriver tip with my air wrench to remove the nut from the

Minneapolis-Moline with a base-mounted American Bosch magneto.

impulse coupling. Remove the washer and pull the impulse cup from the shaft. American Bosch uses the small coiled spring with two balls that ride on a groove within the cup. Set it to one side. Remove the cam bar from the pawl plate. Note the letter on the outside. It should be a C. If the cam plate is stuck to the cup, you will see an A.

American Bosch has always, as far as I know, used sliding pawls in their impulse couplings. Some applications use additional springs to force the pawls down. Most of the time they rely on gravity to cause them to fall. I don't recall ever having to replace a pawl, but I have had to replace a worn pawl stop plate. Note the position of the slot cut into the stop plate. I usually leave the stop plate in place unless everything is really rusty. Four screws hold the stop plate to the main housing.

There are two 1/4in holes in the pawl carrier. These are "fine" threaded holes. I use a pair of 1/4in bolts in one of my two jawed pullers to extract the pawl carrier. Start with some pressure on the screw and give a sharp rap with a light ham-

mer. If it doesn't pop loose, try heating it with a propane torch and rapping the puller with a medium-weight hammer.

Note that the impulse pawl carrier also has a pair of slots for the woodruff key. Again these are labeled either A or C for clockwise or anti-clockwise rotation. For the Minneapolis-Moline tractor we use the C slot.

Coil and Condenser Removal and Testing

Remove the woodruff key. Turn the magneto to the cover side and remove the four screws that hold the plate containing the ball bearing for the distributor rotor and the magnetic rotor. Note how the primary coil wire runs behind the bearing support. Observe the triangular clip under the upper left hand screw. The clip prevents the primary lead from getting tangled with the distributor drive gear. Slide the plate out and remove the magnetic rotor.

Remove the coil retaining screws from the top of the magneto. Slide the coil from the housing. If the screws are too tight, heat the area around the screw with the propane torch (a heat gun also works). Apply pressure with a screwdriver while heating and the screw will almost immediately loosen. Slide the coil out.

Test the coil and condenser with a coil tester, if possible. If not, use the multimeter. The secondary winding should be 8200–9100 ohms. Primary winding tests at 0.6 ohms. The condenser has 0.33 to 0.37 microfarad capacitance. A multimeter set to the Rx10,000 ohm scale should have the needle deflect momentarily when the leads are attached to the condenser. If the meter continues to show resistance of less than infinity, the condenser is leaking. A digital meter will show resistance momentarily before returning to the over range indication.

If you have a source of high voltage, test the high-voltage conductor and the distributor rotor for grounds. If you lack a high-voltage source, use the multimeter and test both parts for continuity to ground.

Cleaning and Examination

At this time I like to examine all the plastic parts for carbon tracking and cracks or flaws. Remove the carbon brushes from the cap by turning them clockwise while pulling gently. Avoid touching the carbon brushes with oily fingers. Short brushes should be replaced. If the cover (part #WN521) is not excessively crusty, try cleaning it with soap and water. If that doesn't do the job use a stronger solvent. Watch out for the window in the center of the cap. Avoid getting solvent on the plastic, if it is still in good condition. If the plastic is excessively beat up or cracked, plan to replace it.

Wipe the distributor rotor with a dry cloth. Use minimal solvent on either the gear or the plastic rotor. If there is excessive wear on the outer groove, it can be chucked in a lathe and turned or sanded. Examine the fiber gear for broken, chipped, or worn teeth. If teeth are missing, be sure to look for the reason. The most likely cause is a defective ball bearing. When in doubt, throw it out.

Clean the housing of grease, rust, and small metal fragments. Look for rust where the coil is seated, and where the magnetic rotor turns. Also, clean rust from the magnetic rotor itself. Remove the two screws from the bottom of the magneto housing. If they are tight, try putting a few drops of penetrating oil where the screws end inside the housing. Use heat if necessary. The holes are used in setting *E* gap, or rather one of the holes is used. When I dismantle an MRF magneto, I always remove both screws. When it is time to adjust the *E* gap, I take the time to figure which

American Bosch MRF4A322 with cover and contact plate removed. The cam has two keyways—one for clockwise and one for counter-clockwise rotation.

one to use. In this case, viewing the magneto from the driven end, the proper screw is the one on the right side.

Pull the Rotor and Bearings

Examine the magnetic rotor and the ball bearings. Make sure that the shaft is straight and the magnet is tight on the shaft. Clean and examine the ball bearings for pitting and scoring. If they are good, let well enough alone. If they must be replaced, remove the bearing cages and balls. Use an inner race puller or a bearing separator and press to remove the race. Use an outer race puller or some device to put a little pressure on the race. Heat the support with a propane torch and ease the race out. American Bosch uses paper bearing shims to insulate and hold the races in place. (Part #52242 is 0.009in thick, and part #52443 is 0.008in thick.)

The oil seal lives behind the outer race at the driven end of the magneto. Feel the lip of the oil seal. Replace a seal that has a hard, unyielding seal lip. If you are to replace the oil seal, you have to pull the outer bearing race. If the oil seal is still flexible and not leaking, appears to be in good

American Bosch MRF series with coil, high tension lead, distributor drive gear, distributor gear, cam, and condenser removed. Everything about this magneto is heavy duty.

shape, and the ball bearings are in good condition, I generally let well enough alone. The seal (American Bosch part #SE3004) is available. (Victor part #46428.)

Reassembly

Observe the spacer washers; if you remove them keep them together. Install them with the new bearings. American Bosch allowed a little more slop than Bendix did. Bosch allows from 0.0005in to 0.002in end play as checked with a dial indicator. Spacer washers come in 0.002, 0.004, 0.007, and 0.012in (part #WA61, #WA106, #WAl07, #WA1009).

Install the magnetic rotor in the housing in the neutral point. Slip the coil into place, with the primary lead to the left and the dimples in the core facing the retaining screws. Slide the insulating fabric into place around the coil. Make sure that the contactors are free in the high-voltage conductor. Install the high-voltage conductor on the bearing plate and install the bearing plate.

Before installing the screws, make sure that the primary lead is run around the bearing plate on the left hand side, so that it cannot get tangled with the distributor gears. Tap the woodruff key in place and install the distributor drive gear with the tooth marked with a C in line with the woodruff key. Place the edge of the washer with the slot aligned with the notch for the woodruff key and the C mark on the distributor gear and slide the distributor gear and the washer into place at the same time (the washer rides between the rotor and the fiber distributor gear and keeps the distributor from sliding out when the magneto is in service).

Install the cam, being careful to engage the notch marked C with the protruding end of the woodruff key. Insert and tighten the cam screw. Install the condenser and the contact plate with the points installed. Attach the condenser wire and the primary coil wire to the moveable contact post.

If the cam stop plate was removed, install it. Install the woodruff key. Look at the reverse side

The American Bosch impulse coupling is one of the simplest and most reliable of all the impulse couplings. Several minor variations of this impulse were used by American Bosch over the years. The direction of rotation can be changed by merely installing the impulse pawl plate so that the C or A keyway slides over the key. The two L-shaped weights are marked C
on one side and A on the other. If the magneto turns clockwise, slide the pawl plate keyway marked C over the woodruff key in the magnetic rotor shaft. The pawls are installed so that the C is up and visible. The cam plate (it looks like a squiggly bar) is installed with the C up. The impulse drive cup works equally well in either direction.

of the pawl carrier plate. Note that there are two slots, one is marked with a *C* the other with an *A*. Since this Minneapolis-Moline tractor calls for clockwise rotation, make sure that the woodruff key goes into the *C* slot. Place the L-shaped pawls in place with the *C* marked side out.

The cam had the letter *C* facing out. If you install it in the slots in the cup, you will see the *A*. I usually use bearing grease to "glue" it into the proper notches in the cup. If the cup is dirty, I like to remove the spring and balls. The easy way is to pluck the spring out with the bent tip of a scribe. The balls can be rolled to the notch in the inner part of the cup and removed.

There is a felt strip inside the coiled spring. Remove and clean in solvent and dry it. Fish it back through with a wire with a small hook at the end. Oil it with SAE 20 oil. (or any handy oil). After drying the felt with paper towels, I usually drop it into my oil-wick jar which I keep on my workbench. When I am ready to reinstall the felt, I remove excess oil and slip it into place. Drop the balls into the slot, with one ball going to the left and the other to the right. Force each end of the spring down on a ball and work the spring back into place.

Pawl Plate Installation

Install the pawl plate with the *C* slot meshing with the woodruff key. Install the cup so the lever goes between the two balls. Wiggle it around until it goes all the way in. The pawl plate axle should be slightly beyond the lip of the cup. Install the washer and special nut. Turn the cup. The pawls should engage the pawl stop, and you should feel the springs exerting pressure. When the impulse

snaps, the magnetic rotor should move freely with no binding at the cup.

Set the points to 0.018in, when the rubbing block is at the high point of the cam. Use a 2mm gap gauge, or a numbered drill to set edge gap. Turn the magneto upside down so that the pawls won't catch. Rotate the magnetic rotor ten or fifteen degrees past the neutral point. Insert the drill or gap gauge into the right hand hole in the bottom of the magneto (while you are facing the contact end). Turn the magnetic rotor back toward the neutral point until the gap gauge or drill catches. Slide the breaker plate until the points just open. If you have a pack of cigarettes lying around, the cellophane from the wrapping works well for adjusting the points. Slip a piece of cellophane between the points. Pull gently on the cellophane while you move the breaker plate. When the cellophane slips out, you are close enough to the proper *E* gap.

Since I don't smoke, I disconnect the coil primary wire from the breaker assembly and use my digital meter with its continuity buzzer to make this adjustment.

Place the magneto in the vise and hold a screwdriver near the high-voltage contactor and turn the impulse until it snaps. You should get a good spark.

Install the cover gasket and the cover and test the magneto in a test bench or on the tractor. I usually leave the breaker plate cover off until I have checked operation. Watch the points for too much fire or shooting stars (shooting stars indicate dirt particles on the points).

If all goes well, install the breaker plate cover and congratulate yourself on a job well done.

Chapter 14

Bosch DU4

The Bosch Model DU4 magneto was manufactured in New York City (prior to 1917) and is my favorite magneto. For one thing, the housing is made of bronze and brass and easily polishes to a golden sheen. For another, every-thing about this magneto shows attention to detail. The cover on the oilers are brass and are engraved with the legend "oil." The direction of rotation arrow is engraved in the driven end oil cover. The front oiler has two funnel-type openings.

The Bosch DU4 was used on a variety of farm tractors and automobiles. The pictured magneto was made prior to World War I in New York City. The housing was made of brass and *steel and all the visible parts have been polished to a bright shine.*

One delivery tube goes to the distributor rotor shaft oil wick, the other tube leads to the ball bearing that lives behind the contact plate.

The DU4 has the high-tension pickup at the driven end. This pickup has two leads. A spring-loaded and insulated high-tension lead conducts the fire to a hole at the rear of the distributor rotor. The second open, uninsulated spring contact touches the top of the safety gap. There is a porcelain cap with a center rivet. Should a lead fall from one of the spark plugs, the spark would jump from the cap to a similar rivet in the body. If gaso-line fumes are present, the fine copper screen prevents the flame from escaping.

A pair of squeeze springs on either side of the distributor cap holds the cap in place. A single rectangular brush transfers the spark from the center of the cap to brass contacts connected to each spark plug wire terminal. With a good coil and condenser, and with the magnet charged, this magneto will produce spark at ridiculously low rpm.

I don't know how many DU4 magnetos were made, but I see several every time I go to a threshing show. The magneto housing and magnets are very rugged and durable. Bronze and brass stand up to severe conditions with smile. The distributor cap is quite rugged, but if they are lost, replacement is difficult. One time a customer came to me with a two-spark Bosch magneto from a Model T Ford with a Frontenac conversion—overhead valves, two spark plugs per cylinder, and all kinds of go-fast goodies. The distributor cap on this magneto was defective. I called people from the east coast to the west coast. Finally, I found I fellow with a completely rebuilt magneto with a good cap. Naturally, he wouldn't sell me just the cap. "A thousand bucks will get you the whole thing," he said. My customer passed. Later, he found a good used magneto for three or four hundred.

This magneto was last used to fire an automobile engine. It did not use an impulse coupling but was driven by a simple coupling. I would hate to hand crank an automobile with no impulse in cold weather. I repaired a six-cylinder Bosch magneto that had an extra cam and ignition contacts as well as a distributor cap with two rows of wires. Each cylinder had two spark plugs. It was started on a battery and ran on both ignition systems. The car was a 1930 Mercedes-Benz type S and had an aluminum-block, single overhead camshaft, seven-liter engine.

The DU4 magneto has patent dates of October 17, 1905, October 24, 1905, and October 6, 1908. Other than the neat features mentioned, the magneto is a straightforward shuttle-wound magneto.

Disassembly

The first item to be removed and placed in a safe place is the distributor cap. Squeeze the two spring clips and pop the cap off. Examine for wear and carbon tracking. I always remove the carbon brush from the distributor rotor at this time and keep it with the cap. The high tension lead has a spring-loaded clip, and can be removed from front or rear. Wipe it with a dry rag and place it with the cover for safekeeping.

Remove the impulse coupling, if there is one. Remove the woodruff key from the shaft. Since you are working on the drive end, now is a good time to remove the metal clip that holds the safety gap ceramic cover and put it away. A knurled nut holds the high tension pickup in place. Unscrew the nut and carefully remove the pickup. Make sure that the carbon brush is in place and in good

The best way to store Bosch magnets is to turn them one hundred-eighty degrees and stick them together. Single magnets require an iron or steel keeper. Magnets left lying around pick up sharp steel filings and miscellaneous small parts while they lose magnetism.

A DU4 with the pickup, high-tension lead, and safety gap cover and magnets removed. Every part shows quality and engineering.

condition, and place it into the vault with the other irreplaceable parts (I guess I am pretty paranoid about fragile parts from very old magnetos).

Turn the magneto over and slip the spring steel brush retainer to one side and remove the high copper content ground brush. Remove the four screws on each side of the magneto and remove the two horseshoe magnets. Instead of a keeper, rotate one magnet one hundred eighty degrees and stick it to the other magnet.

A DU4 armature with a special wrench to remove the nut at the center of the distributor gear (behind the inner race). The new coil has extra linen string wrapped around the outside of the coil armature and is dipped in varnish and baked. The condenser lives in space between the coil and gear. The special wrench has been around my shop forever. I don't know where I could get a replacement. A machinist could make one for a price.

Remove the contact plate retaining screw. The contact plate will usually slip off at this time. Remove the cam housing. If the contact plate is still stuck in place, you can pry it loose with a pair of small screwdrivers. Remove the contact cover retainer bolt and the two screws that hold the triangular bearing plate to the housing. Remove the coil armature assemble and the distributor gear at the same time. Note the C and A marking on the brass gear. The gear on the armature has a beveled tooth.

Test the Coil

If the coil is soft or has leaked black tar, it must be replaced. In any case, I always unsolder the wire that is attached to the insulated side of the condenser. Test the condenser and the coil with a coil tester or with your multimeter. Make sure that there is continuity between the collector ring and the high tension winding. If you show 7000–8000 ohm resistance from the primary lead and the collector ring, the coil is probably okay. I'm never happy until the coil has passed the test on the old Eisemann coil tester after a half hour of constant running.

Test the Condenser

Test the condenser for capacitance, leakage, and series resistance. I have had good luck with mylar insulated, axial lead, two or three hundred volt, 0.1 microfarad capacitors from the local electronics store. I have used capacitors with 0.2 to 0.4 microfarad capacitance also with good results. If the points don't spark very much under operation of the magneto, the capacitor is the right size. A tiny blue spark is all you should see.

The Armature

If the coil or condenser fail the tests, the armature must be taken apart. The inner races may be

The bottom view of the DU4. The ground brush is visible at the center of left side of the base. A flat spring is supposed to be attached to the screw at the center of the X in the base. It holds pressure against the ground brush.

removed with a bearing separator or special puller. The gear is held in place with a notched nut which should be removed with a special or spanner wrench. If worse comes to worse, the ring can be turned with a punch and a small hammer. Once the gear has been removed, you can unscrew the flat head machine screws that hold the cap to the coil. Make sure that the screw driver blade fits exactly. These screws are customarily staked in place. The metal moved into slot can be removed with a sharp, small screwdriver. Always remember that the brass parts are easily bent. Handle with care! With the bearing race removed, the spacers and the collector ring may be removed. Hold the collector ring as best you can and, with the nut installed on the driven end to

protect the threads, tap the nut with a soft or brass hammer.

Carefully clean the varnish from the ends of the new or rewound coil where it fits into brass end pieces. If you have been careless in your handling of the armature, now you pay! If the end pieces have been distorted, or if the varnish on the coil prevents the end pieces from fitting tightly, you may find that the armature now rubs in the housing. If this is the case, go back to the beginning and start over. As a last resort, the armature may be trued in a lathe. This will not improve performance.

Install the distributor drive gear with its retaining nut. Replace any bad bearings. Re-use the spacers if possible. End play should be from

Details of the rotor and distributor cap. High voltage electricity is conducted by the lead that connects the pickup at the driven end to the center of the distributor cap. An insulated block is fastened to the bronze distributor gear. The block has a spring-loaded carbon brush that bridges the gap between the center ring of the distributor cap and the four brass strips around the edge of the center depression. The brass strips are connected to the spark plug wires. I hate to think of how much it would cost today to reproduce this magneto.

The distributor gear can be removed at the same time as the armature. I always install them together after matching the timing marks.

The dark spot in the center of the bushing is an oil wick. Lubricating oil runs down a tube into a reservoir.

0.0005 preload to 0.0015 play. Fill the spaces between the balls about half full of good bearing grease.

Slide the armature in place with the distributor gear at the timing marks and install the bearing retainer plate, points, cam ring, and cover. The safety spark cap and the high tension pickup may be installed with the high tension lead and the distributor brush and cap. Remagnetize the magnets, if possible, and install them. The woodruff key, impulse assembly, and retaining nut may be attached. Don't forget to replace the ground brush and tighten the spring steel retainer in position to press against the brush. Test it on tractor or test bench.

Chapter 15

Eisemann G4 1928 (Aultman-Taylor)

In August, 1993, I attended the Threshing Show at Jordan, Minnesota. I was walking around snapping pictures and swapping lies with some of the old duffers who hang out at these affairs. I ran into one of my customers and, with my unerring skill at starting conversations, I led off with:

"Hey, Gary, I have just seen the funniest tractor in the world. It's parked over there behind the big Minneapolis tractor. It looks like the builder changed his mind half way through building a steam tractor. They kept the boiler, fire tubes, and all, but—get this—they laid an unfinished four-

The Aultman-Taylor running with gears, clutches, chains, and sprockets hard at work.

A 1928 Aultman-Taylor tractor. This tractor used the Model G Eisemann magneto. Another Aultman-Taylor at the Jordan Threshing Show used an HW Magneto. I wouldn't be surprised to find a Bosch DU4 on another Aultman Taylor.

cylinder engine on its side and called it..." I noticed a little frost forming on the face of my friend, so I stopped talking.

"..A 1928 Aultman-Taylor," he said. Then he smiled. "The radiator holds one hundred twenty-five gallons of water," he said. "Back then, tractors left lots of parts exposed."

All in all, I was a little embarrassed to have made unkind remarks about what was actually an interesting tractor. I climbed on board and took some pictures and found out where the magneto was hidden. "An Eisemann G Series," I exclaimed, "I haven't seen one of those in years."

Well, I hadn't. The next week I was visiting the Threshing Show at Rogers, Minnesota, when I came upon one of my most favorite vehicles, an old Mack truck. The hood was open, and sure enough, there was the old Eisemann G Series in all its glory. (I saw a documentary film when I was in grade school about the construction of Bolder Dam on the Colorado River. The Mack Truck hauled a lot of dirt in its day, which lasted into the fifties.)

Eisemann, like Case Magneto Company, produced a full line of magnetos for everything from tractors to light aircraft. Eisemann produced a shuttle-wound coil type magneto, a well-designed inductor magneto and an excellent rotating mag-

A better view of the Eisemann G4 magneto. I had overhauled this magneto two days before this picture was taken. During the first day of the Threshing Show, the owner had started the tractor with hand crank. The next day, when I was at the show, the owner chinned himself on the crank about forty

times without a single pop. I was embarrassed. Later he used another tractor with a flat belt to spin it over and it started and ran just fine. The magneto is back in my shop, and I will run it on my test bench to determine if I was at fault.

An Eisemann G in a Mack truck.

net type which had wide spread industrial applications. My first airplane had a pair of Eisemann LA4 magnetos which never dropped a spark, except at night when flying over large bodies of water. Of course, all engines *seem* to run rough at night over water. Pilots call this jittery sensation "automatic rough."

The Eisemann G has swing-away spring cap and cover retainers, which is how I identify the magneto from a distance. Otherwise, the G is a straight forward shuttle-wound magneto.

Disassembly

As always, remove the cap and cover and put them to one side for safe keeping. Check the cover for carbon tracking. Remove the brushes for safe keeping and don't forget the pickup brush at the bottom of the cover. Remove the magnet and place it on a metal keeper. Remove the safety gap screws around the perimeter of the cap housing. Take the contact plate screw from its socket and remove the contact plate.

Remove the three screws that hold the cam ring and bearing plate and set them aside.

Remove the impulse coupling. Extract the woodruff key. The armature may be pushed forward and out of the housing at the same time the distributor plate is removed.

Unsolder the insulated lead to the condenser and test the condenser for shorts, capacitance, leakage, and series resistance. Test the coil on a coil tester or with a multimeter. The secondary should have about 7000 ohms resistance, the primary 0.3–0.4 ohms.

If the coil must be replaced, remove the inner race at the contact plate end. Remove the collector ring and gear. Remove the four screws that hold the end cap in place. Remove the four screws at the driven end and gently separate the end caps from the coil. Be sure to remove excess varnish from the ends of the coil and clean the inner surface of the end caps. Test the condenser. It can be replaced by a mylar insulated, axial lead, 200–300 volt, 0.1–0.2 microfarad capacitor. If I have time to spare, I like to install the condenser in the original condenser can. Be very careful when unsoldering the seams of the condenser. These old condensers were usually sealed with beeswax which can catch on fire, or which when melted tends to stick to tender flesh. When in doubt, don't try to dismantle the old condenser. Instead, you can just secure the condenser in the proper place with a drop of super glue gel or epoxy. You can fill the recess with silicon sealer.

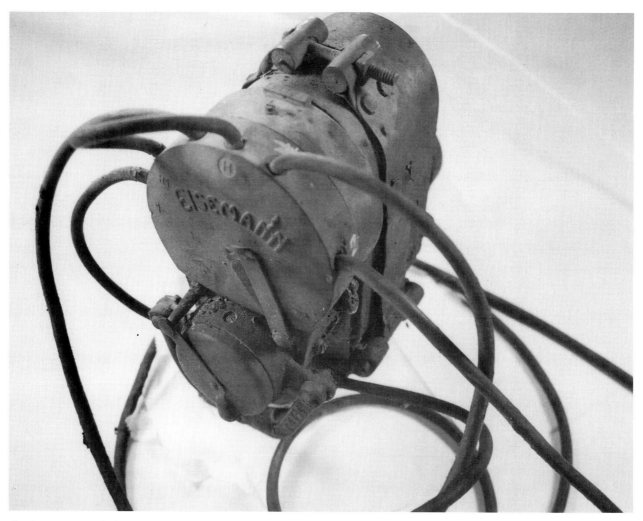

The Eisemann G with the mounting bracket attached. Ten seconds with a screwdriver and the magneto can be removed from the engine for repair.

The housing stripped for cleaning.

Detail of the cap. Note the jumper that conducts the spark to the center contact. I haven't seen this setup on any other magneto.

When you have completed the installation, install the armature in the housing and make sure that it turns freely. If it rubs, take it apart and try again. If you cannot get it to work any other way, chuck it in the lathe and remove the high spots. Remove a minimum of material.

Install new bearings or, if the bearings are okay, pack the old bearings with bearing grease and install the armature and distributor gear in the housing. Install the contact plate. Magnetize the magnet, if possible, and install it. Clean and lubricate the impulse coupling and tighten the retainer cup washer, lock washer, and nut. Check points and set to 0.012in. Install the safety gap screws. Install the cap and cover and test on test bench or tractor.

Chapter 16

Electrical Systems

The electrical system of your old tractor consists of the following components: battery, generator, regulator (or cutout), starter, ammeter, switches, lights, the wiring harness, and ground (the iron block and steel frame).

Generator

Generators come in a number of variations. They may be totally enclosed (sealed) or air cooled. They may have six-, twelve-, or twenty-four-volt output. Some generators turn clockwise; some turn counter-clockwise. Output of a generator may be controlled by a voltage regulator or by using a "third brush," a field resistor, and a cutout. They may be gear-driven or belt-driven. Generators are simple devices and most can be repaired at a reasonable cost.

The generator on your old tractor may have been manufactured by a number of different manufacturers. The most common were those made by Delco-Remy and Autolite. In my experience, domestic and foreign generators are all very similar. They consist of a field frame which contains the field coils and field coil pole shoes. A plate at the driven end contains a ball bearing. Outside the driven end plate are spacers, a fan (if the generator is air-cooled), pulley, lock washer, and nut. The other end is the commutator end (com end). The com end plate contains a bearing and, frequently, the brush holder assembly. Generators usually use a brass or porous bronze sleeve (Oilite) bearing at the com end. Some generators use a ball bearing at the com end.

The Armature

The most important part of the generator is the armature. Electricity is generated within the armature and transferred through carbon brushes and wires to the rest of the system. The armature consists of a steel shaft run through a stack of flat stampings. Coils of wire wind through the notches in the stack and attach the commutator.

The wire used in the armature coils is called magnet wire. It has a solid copper core with a coat of varnish to insulate it electrically. The commutator consists of copper blocks arranged in a circle around a steel core. Each bar is insulated from the next bar and from the shaft by mica insulators. The windings are insulated from the stack by a special insulating paper and may be wedged in place with wooden wedges.

After the armature is wound, the varnish is scraped from the ends of the wires. The wires are then soldered to the proper commutator bar. A wire starts at a commutator bar, wraps five or six turns and ends at the next commutator bar. The next coil starts at that bar, wraps five or six turns and ends at the next bar. This continues all the way around the armature. Finally the armatures is dipped into hot varnish and baked. The varnish locks the wires securely in place and keeps the wires from chafing each other or the wearing through the insulation to ground.

Brush holders are usually, but not always, part of the com end plate. One brush holder is insulated from the plate: The other is grounded directly to the plate. In practice, the brush holder closest to the terminals is the insulated holder. I always remove the cover band and disconnect the wires from the brush holder and remove the brushes.

Generator Disassembly

At this time I like to insert one finger from each hand into slot so that I can exert pressure on the commutator and determine wear in the com end plate bushing or bearing. If there is appreciable side to side movement of the armature, the bushing or bearing will have to be replaced.

Remove the through-bolts next and gently pry the com end plate loose. If you are not familiar with the generator, and if there is an external plate over the bearing, it is wise to remove the cover and make sure that there is not a ball bearing fastened to the end of the shaft.

End play is usually controlled at the driven end. The outer race of the ball bearing is held in the end plate by a ridge at the outer edge and by a retainer on the inside. The inner race butts against a shoulder on the inner end of the shaft. A spacer rides against the inner race and is held by the pulley, nut, and lock washer.

Once the com end plate is out of the way, rap the driven end plate with a brass or soft hammer. Once the plate is loose, slide the armature from the field frame.

Hold the armature in the vise (use soft jaws), and remove the nut and lock washer. If the pulley is cast-iron, use a puller to remove it. If it is a stamped part, use the puller gingerly as the metal has a tendency to bend. A ventilated generator has openings in the plate. You may use a brass drift and hammer to drive the pulley from the shaft. I use a drift with a beveled tip and hit as close to the center of the pulley as I can get. Heat or penetrating oil may help to free a stuck pulley. Dents and bends can usually be hammered straight.

After removing the pulley, remove the woodruff key and the spacer. The armature may be pressed from the driven end plate (or use a three-jawed puller.) Remove the bearing retainer and press the ball bearing from the plate.

Before cleaning the field frame, examine the inside of the frame for signs of thrown solder. If you see a ring of light gray or silver at the com end, take another look at the armature where the wires are soldered to the commutator. If dark voids appear where solder usually appears, solder the armature.

Testing Your Generator

If everything about the armature looks all right, test for opens, shorts, and grounds using a growler. Set the armature in the growler, turn on the current and hold an old hacksaw blade lengthwise on the top of the armature. Turn the armature with the blade nearly touching at the top. You should not feel any magnetic attraction of the blade to the armature. A defective armature will cause the blade to vibrate, indicating short circuits between the windings on the armature. Turn off the growler and test for grounds with a 110 volt bulb. Touch one prod to the commutator and touch the other to the shaft. If the light comes on, the armature is grounded.

Finally, test for open circuits in the windings. If the growler is equipped with a meter, follow the instructions. The readings should be very close. A defective winding will give a lower reading. If the growler is not equipped with a meter, turn the growler on and use the hacksaw blade to short circuit bars together at the ninety degree from vertical position on the commutator. Sparks should fly when two bars are shorted. Turn the armature and short circuit each pair in turn. The amount of sparking should be the same bar to bar all the way around. If one pair of bars doesn't spark, or sparks weakly, an open winding is indicated. I like to confirm my finding with a sensitive multimeter. Sometimes you can save the armature by soldering the connections at the commutator. If a wire is broken the armature must be replaced or rewound.

Brushes

If the brushes are worn, they should be replaced and the armature's commutator should be turned true in a lathe. By the time that the brushes have worn to a nubbin, the commutator will be worn egg shaped. New brushes will last much longer if the commutator is ground to a true circle.

Mica between the commutator bars should be undercut about 3/64in. An undercutting tool can be made from an old hacksaw blade. I always flatten the teeth and, if necessary, narrow the blade with the bench grinder to achieve the proper cutting width. A wooden handle or a few wraps of tape help to protect the fingers. Recheck the armature on the growler after undercutting (sometimes, a thin sliver of copper will bridge the gap between bars, effectively short circuiting the armature). Finally, the commutator should be sanded with fine sandpaper and polished.

Field Coils

Examine the field coils. They are usually wrapped with cotton, linen, or paper tape. If the wires are visible through tattered tape, the fields should be removed from the field frame and rewrapped. If the varnish insulation of the wire is very dark or if it flakes when you touch it with a finger nail, the field coils should be replaced. If the fields appear to be usable, measure resistance with your multimeter. As a rule of thumb, six-volt fields will have two to three ohms of resistance; twelve-volt fields will measure between five to eight ohms; twenty-four-volt fields have eleven to twenty ohms of resistance. If specifications are available, use them.

Removal of pole shoe screws usually requires an impact driver with the correct bit and a large hammer. Wallop the frame with the large hammer a few good blows. The shock may help to loosen the screws. Upon installation, use an impact driver to tighten the screws as much as possible.

Bearings

Ball bearings may be cleaned by washing in solvent, rinsing with hot water, and blowing dry with compressed air (do not spin the bearing with compressed air). Examine the balls and races; replace them if any checking or pitting is found. Pack them with a small quantity of ball bearing grease. Fill the available space about one quarter full. Clean the felt washers and reinstall the bearings. If the outer race has been turning in the driven end housing, knurl or punch a series of marks around the circumference of the housing. Loctite the bearing in place. Most of the old tractor generators used a 203 double-sealed bearing.

Sleeve bearings present a little more of a problem. Many standard bushings are still available through rebuilder suppliers like ACE or IPM. They can be found either by part number or by dimension. A friendly small rebuilder can be of great help in locating such items. Many are available in undersized versions to compensate for a worn shaft. If you can't locate one, they can be manufactured on a lathe from either porous bronze or brass.

Brushes

If possible, brushes should be replaced. If the exact part number is not available, brushes can be matched or a larger brush cut down until it fits. It is important to try to match brushes according to composition. Brushes are made with varying quantities of carbon, graphite, and copper. First match color. A brush can be a soft black or a hard black or may be a copper color. Feel the brush. Brushes with graphite feel slippery to the touch

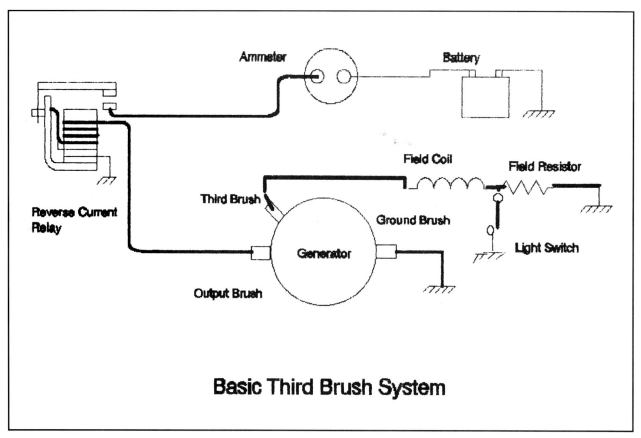

Basic Third Brush System

The basic third brush system is simple, elegant, and effective. A regulation system that is current sensitive has a certain advantage over voltage sensitive regulation when operating under severe conditions. I had a customer, many years ago, who operated a forklift truck in foundry under rather corrosive conditions. His previous forklifts had used an Autolite or Prestolite GAS series third brush generator. He had few problems with this setup. His new forklift had a Prestolite alternator. After a month or two the battery would run down. He sent the alternator to us for warranty repair. There was no continuity through the brushes and slip rings. We cleaned the slip rings and replaced the brushes and sent it back. Within the month it was back with the same problem. Under the same conditions the third brush generator operated flawlessly. Two factors contributed to this success: The pressure of the three brushes on the commutator bars kept them clean and the rise of voltage until regulated current was reached burned away corrosion on mechanical electrical connections. We finally solved the problem by having the customer install several bright lights which were turned on whenever the forklift was operating. The slight increase of current through the slip rings, along with the use of brushes with a high copper contact, kept the alternator operating.

A typical third brush generator. The cover band has been removed. While this generator came off of a tractor, it is a ventilated or air cooled generator. Most tractor generators are non-ventilated or sealed to keep the dirt out of the works.

when compared to a pure carbon brush. If there is a wire embedded in the brush, try to find a brush with the same or larger gauge wire. Sometimes you luck out and find a perfect match. Other times you have to make do. Brushes are carefully matched to the application by the design engineer—voltage, rpm, load, spring pressure, and operating temperature are taken into consideration. Depending upon the degree of mismatch, brush life will be affected. Fortunately, antique tractors are not expected to spend long hours plowing the back forty.

To seat the brushes, install them in the brush holders, wrap a piece of fine sandpaper around the armature, insert the armature in the end plate, and rotate the armature several times or until the brushes closely fit the contour of the commutator.

Install the armature in the driven end hous-

ing. Install the spacer and fan, if required. Install the pulley, lock washer, and nut. Tighten the nut securely to sixty or seventy foot-pounds. The woodruff key is there to aid in nut removal, if you don't wish to dismantle the whole generator when you replace the pulley. If the nut is not tight enough, the key will wear both the shaft and the pulley, rendering both useless.

Slip the armature into the field frame assembly. Install the com end frame. Note that there are dowel pins and holes, or a rivet with a round head that slips into a groove. It is important that the parts are aligned. Install the through bolts and tighten. Install the brushes. Make sure that the wires are installed properly. Given half a chance, the field or output wire will rub on the armature. I like to test the generator before I install the cover band. That way proper functioning of the brushes can be observed.

During operation, a tiny blue line of sparks should visible where the brush contacts the commutator. Excessive arcing indicates that something is wrong. It could be a defective armature, a piece of high mica, or a weak brush spring. Specifications indicate the proper tension measured with a spring scale. If you haven't the specifications, compare tension between the two brush holders. Improper tension will cause poor operation or short brush life.

The Third Brush Generator

Many antique tractors used generators with a third brush. The third brush is used to regulate output. In order to deal with third brushes, it helps to review how a generator works.

We start with a little left over magnetism in the pole shoes. This is called residual magnetism. When a conductor is moved through a magnetic field, voltage is generated. When the conductor completes a circuit, current will flow. Since the voltage is less than battery voltage, the cutout is open so that the battery is not connected to the generator.

The third brush conducts current from the armature to the field coils. It travels through the field coil to ground, where it returns to the armature through the ground brush. When current flows through a conductor, it creates a magnetic

The third brush is the brush closest to the terminal at the top of the generator. The output brush is next to it. The direction of rotation is always from the third brush to the output or ground brush (whichever is closest). This knowledge—that the third brush is closest to the output brush—means that the voltage regulator, if it has one, must be an A circuit (regulates to ground) regulator. If the third brush is next to the ground brush, the regulator would have to be a B circuit.

field. Since the conductor is in the form of coils surrounding cores of soft iron, the resultant magnetic field is concentrated between the pole shoes. The armature turns within the stronger magnetic field. The voltage rises. When the voltage at the output brush terminal reaches six volts, the cutout (reverse current relay) connects the generator to the battery. The voltage continues to rise, and the current charges the battery.

The third brush is located one or two commutator bars in front of (leading) the main output brush in direction of rotation. The reason for this location becomes clear when we examine how magnetic fields interact.

As the armature turns within the magnetic field, electricity is generated in the rotating coils of the armature. The current flowing in the armature wires generates a magnetic field in opposition to the magnetic field formed by the field coils. Since the armature is rotating, the field is distorted in the direction of rotation.

If you look at an armature from the com end, with the ground brush at the bottom and the output brush at the top, current flows from the bottom brush to the output brush. The current splits, with half flowing on each side of the armature. Actually, current flows from the bottom brush through the coil to the next commutator bar. If you measure voltage between the ground brush and half way to the top, you find that the voltage is about half the voltage measured from ground to output brush.

The voltage at the third brush is, therefore, less than the voltage at the output brush. For the sake of discussion, say that the voltage at the output brush is seven volts, while the voltage at the third brush is six volts. When the generator speeds up, the voltage increases. Ah-ha! The increase in current flow causes the magnetic field to be distorted in direction of rotation. The voltage at the third brush position decreases slightly. This reduces the flow in the field winding which weakens the magnetic field and lowers the output of the generator. Less current means less distortion of the magnetic field, which in turn allows a voltage increase, and the cycle runs again.

While the theoretical process is one of constant flux, the real-world result is a constant flow of electrical power over a range of speeds. Output drops off at low and high rpm; the exact range varies by generator model and the position of the third brush.

Adjusting the Third-Brush Generator

This kind of tractor generator is known as a moveable third-brush generator. By changing the position of the third brush, the output may be regulated to the individual requirement. The prop-

er way to adjust a third-brush generator is to use a battery hydrometer and the vehicle ammeter. The output should be adjusted so that the battery is kept fully charged but doesn't use an excessive quantity of water. Begin by moving the third brush toward the output brush until the vehicle ammeter reads three or four amps. After the generator has been rebuilt, the amount of charge indicated by the ammeter can be adjusted. Check the battery cells with the hydrometer on a regular basis to determine the trend. If the generator is overcharging the battery (as indicated by excessive water consumption), move the third brush away from the output brush by one bar. Regularly test the battery and make adjustments until the battery maintains a full charge—as indicated by the hydrometer—but doesn't use excessive water.

Third-brush regulation is current sensitive. If you set the third brush so that the output of the generator is ten amperes, the generator will try its best to put out ten amperes. If the battery is not in the circuit, the voltage will rise. If there is no other path for the current to follow, excessive electricity will be forced through the field coils which will overheat and fail. Old tractors with magneto ignition may be started and run without the battery installed. If you don't have a battery, you can operate safely if you remove the third brush or if you install a jumper from the output terminal on the generator to ground. You can also install a resistor from the battery terminal on the cutout (reverse current delay) to ground.

Barbwire Regulation

When I was a maintenance officer in the Army, a farmer-type mechanic told me that he used a three-foot length of barbwire in place of the battery. He described how the wire would get red hot as the day went on. "Man, if you take the third brush out of the generator when you're in the back forty, you lose it before you get back to the shed," he said. "Then what will you do? Go to town? No, indeed. You just hook up about three foot of barbwire and keep that generator happy and working. Besides that, you have a ready-made cigarette lighter right at hand."

At the time, I thought such treatment would ruin the generator. With the passage of years, I realized that although this solution was not in any book, it was actually a sound procedure.

Varying Loads

Using third-brush regulation presents an additional problem where varying loads are used. If there is a regular pattern of use, the third brush may be adjusted to an average setting. In the golden days of my youth, I drove a Model A Ford with a third-brush generator. As a penniless student,

Commutator end plate with fixed third brush. The hot and ground brush are always one hundred eighty degrees from each other. The third brush is somewhere in between.

the battery in my Model A Ford was marginal at best. If I drove for an extended period on the highway, I either had to get out and adjust the third brush for reduced output or drive with the headlights on. Since I had installed sealed-beam headlights in my car, the generator could barely keep up with the load at night. I got to be very good at changing the adjustment of the third brush.

Many tractor manufacturers were more sophisticated than the car people. They attached a resistor to the light switch that was in circuit when the lights were off and bypassed to ground when the lights were on. This resistor was attached to field coil at the end opposite to the third brush. With the lights off, the output was reduced. With the lights on, the output compensated for the extra load. Actually, my 1931 Chevrolet had a light switch-operated resistor.

The next step was the development of a step voltage regulator which also incorporated the reverse current relay into one unit. Initially, the generator charged at the maximum rate. When battery voltage rose to a certain setting, the points opened, reducing the charge rate. After a time, when the load caused the battery voltage to decline, the contact points closed, increasing the output. The third brush was retained to limit the maximum output to a safe point.

Vibrating-Contact Voltage Regulator

Eventually, the vibrating contact voltage regulator was developed. This voltage regulator consisted of a reverse current relay and a voltage control which maintained a constant voltage. The vibrating contact regulator switched off and on much more rapidly—about one hundred to one hundred fifty times a second—than the step regulator. Field coils, since they are coils, tend to resist current changes. As a consequence, the flow of current through the field coil is fairly constant and produces a consistent magnetic field, which in turn produces an output based upon demand. The aim is to maintain the voltage at a point which maintained the battery at full charge but didn't cause excessive use of water.

My 1937 Buick used a two-unit vibrating contact voltage regulator but retained the third brush to limit current. In this instance the third brush was not adjustable.

The next step in the development of voltage regulators was the three-unit vibrating contact regulator. The third unit was a current-sensitive relay. The contacts of the voltage control relay and the current-limiting relay are hooked in series. When either pair of contacts is open, the field current must go to ground through the resistor. The contacts of the current limiter operate at about half the rate of the voltage control. If you attach an oscilloscope to the field terminal of an operating voltage regulator, the difference in rate of operation is quite visible. The pattern is a square wave.

A or B?

Voltage regulators use either an *A* circuit or a *B* circuit. In an *A* circuit generator, one end of the field coil is attached to the output brush and the other is attached to the field terminal. In a *B* circuit, one end of a field coil is grounded to the frame and the other end terminates at the *F* terminal. An *A* circuit regulates field coil to ground; a *B* circuit regulates current to the field coils.

The *A* circuit or *B* circuit description of generator charging circuits is universal. It includes modern alternator systems in the US and everywhere else. When you look at a voltage regulator or a generator, one of the first questions to be answered is. "Is it *A* or *B* circuit?' The easiest way to find out is to measure resistance with an multimeter. Unhook the unit and measure the resistance between the *F* terminal and the regulator base. If resistance (Rx1 scale) is zero, the regulator is *A* circuit. If resistance is twenty or thirty ohms, measure between *F* terminal and *A* or *Gen* terminal. If resistance is zero then the voltage regulator is *B*

circuit (if reading is also twenty or thirty ohms, you have a defective voltage regulator of unknown type).

If the circuitry of a generator is unknown, first check for a third brush. If a third brush is present along with an *F* or field terminal, check to see which brush is closest to the third brush. If the third brush is next to an output brush, the generator is *A* circuit. If the third brush is next to the ground brush, the generator is *B* circuit. Note that the direction of rotation is always from the third brush to the main brush.

If the generator is a two-brush generator, remove one of the brushes and measure resistance between the armature terminal and the *F* terminal. If the resistance is from two to ten ohms, the generator is an *A* circuit generator. If you find infinite resistance, the generator is probably a *B* circuit generator. Confirm by measuring from *F* terminal to ground. It should read from two to ten ohms (about twenty ohm higher if it is a twenty-four volt generator).

You cannot use an *A* circuit regulator with a *B* circuit generator or vice-versa.

Whenever you have worked on a generator or regulator or battery, polarize the generator before you start the engine. A short jumper wire or even a screwdriver is all that is required. After everything is hooked up and you are ready to start the engine, momentarily touch the jumper wire from *A* (Armature) terminal on the regulator to the *Bat* terminal. Do this two or three times. The spark should hiss when you disconnect the wire. It should not crack and shower you with molten copper. If it does, something is wrong. Never touch a hot wire to the *F* terminal on an *A* circuit regulator.

Jumping *A* to *Bat* will work on either *A* or *B* circuit systems. If you have a *B* circuit system, and are sure that you have *B* circuit system, you may run the jumper from *Bat* to *Field* momentarily.

Commonly, the voltage regulator makers will have the *Bat* terminal next to the *A* or *Gen* terminal on *A* circuit regulators. They also place the *F* terminal next to the *Bat* terminal on *B* circuit systems.

Generally, Ford used *B* circuit systems. Delco systems are usually *A* circuit. But you can't tell unless you check. Take nothing for granted!

Chapter 17

Voltage Regulator Field Repair & Testing

Your neighbor asks you to check over her modernized Earthworm tractor for an electrical problem. Her neighbor told her that you had read *How to Restore and Repair Your Tractor Magneto*. She considers you an expert.

You observe that the engine starts and runs fine courtesy of the Dixie magneto that you repaired last week. You set the hand throttle to the fast idle position. However, the ammeter needle points to zero unless the lights are turned on. Then it shows ten amps on the discharge side of the gauge. You look at the wire harness with dismay. It looks like a rat's nest. "Oh, well," you sigh, "it probably looks worse that it really is." You fetch your multimeter, and several jumper wires.

You note the presence of a six-volt battery and set the meter to the ten volt direct current scale. This tractor has positive ground so you attach the red lead to the voltage regulator base. The black lead is hooked to the *Gen*, *Arm*, or *A* terminal. (Some manufacturers call the center terminal *Generator* others call it the *Armature* terminal. No matter, it is the same thing.)

The other end of that wire is hooked to the *Armature* terminal on the generator. Start the engine and look at the volt meter. It reads one volt. The generator and voltage regulator appears to be a common two-brush Delco-Remy system. The regulator probably uses an *A* circuit. So you disconnect the wire from the *F* terminal and ground it against the base of the voltage regulator. The sound of the engine changes slightly, and the ammeter shows a charge rate of fifteen amps. The volt meter shows seven volts.

Your neighbor asks you, "What gives?"

"Well," you answer, hooking your thumbs in the armholes of your vest and shifting the bit of hay from one side of your mouth to the other, "the generator is all right, as is the harness from the generator to the regulator. The regulator base is properly grounded. The reverse current relay is functioning properly. The voltage control isn't

working. Probably has dirty points."

Your neighbor is impressed. You turn off the engine and disconnect the voltage regulator from the system, having first disconnected the ground cable at the battery. You then set the multimeter to the Rx1 scale and measure from *F* to ground. The reading is twenty-two ohms. To be on the safe side you measure from *Gen* to *F* and note a reading of thirty-five ohms. The first reading confirms your suspicions that the points are not making a clean connection. The second tells you that the voltage regulator is probably an *A* circuit regulator. In any case, it must be cleaned up. (When you have taken it apart be sure to check. Don't laugh! Mismatches occur just often enough to drive you nuts.)

One time I had a good customer who sent in a generator for a clean up and test. It was set up as an *A* circuit. Later he sent in a voltage regulator to be tested and set. The regulator was a *B* circuit regulator. No one told me it was from the same tractor, and I didn't ask. I checked, cleaned, and adjusted it and returned it to the customer. The poor customer tried everything, but couldn't make the two units work together. One Saturday morning I went out to see him and soon realized that we had a mismatch. A *B* circuit Voltage regulator definitely will not work with an *A* circuit generator. At this point he went to an old Ford he was restoring and found the right regulator. "I wondered why the bolt holes didn't quite match," he said.

Field Testing

Adjusting voltage regulators in a well-equipped shop is a snap. Doing it in the field is something else. To do the job according to the book, you must have a good voltmeter and a good ammeter as well as some device, such as a heavy duty carbon pile or a bank of lights, to provide the proper load. If you have a piece of test equipment specifically designed for the purpose of testing electrical systems, follow the directions.

If you lack almost all of the test equipment, charge up the battery and test with a hydrometer. If it has a weak cell, substitute a fully-charged battery. If you can find a thirty-watt, 0.25 ohm (for a twelve-volt system) or 0.125 ohm (for a six-volt system) resistor, you can get by with a defective battery.

Remove the regulator cover and gasket and place them to one side. Hook the voltmeter between the armature terminal on the voltage regulator and ground. Install the test ammeter between the *Bat* terminal on the regulator and the wire leading to the tractor's ammeter. Start the engine and observe the voltmeter. On a twelve-volt system it should be approximately fourteen volts.

Six-volt systems should read about seven volts. The ammeter should say about six amps.

Allow the system to warm up to normal operating temperature. After operating for ten or fifteen minutes, the base of the voltage regulator should be very warm to the touch. Place a thermometer close to the regulator and note the ambient temperature. It is important to be aware of the temperature since most voltage regulators are temperature compensated. The lower the temperature, the higher the voltage.

How They Work

Before trying to make adjustments, we must consider how the parts of the regulator work. This regulator is a three-unit, vibrating contact, *A* cir-

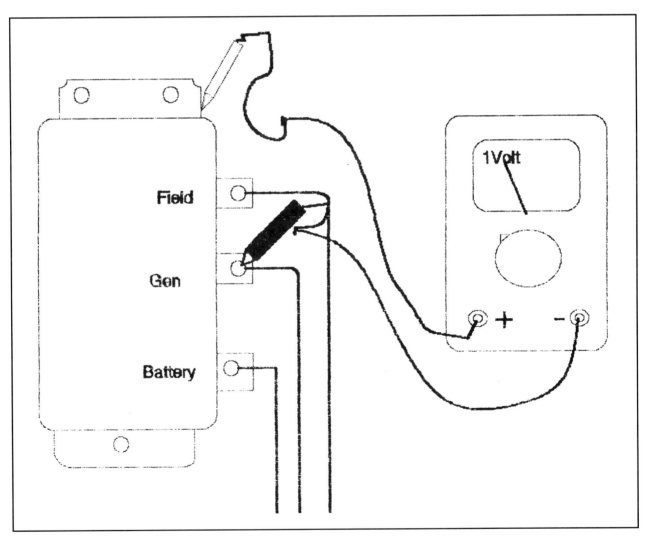

A multimeter in action. In 1956, I bought my first multimeter. It cost ten dollars and was of doubtful accuracy. I measured every electrical thing that I could get my hands on. Gradually, I burned out just about every function, the ammeter section first, ohmmeter section, and alternating current voltage. I kept throwing it into the trash and people kept bringing it back. I think I still have it.

cuit, designed for a positive ground operation.

The first relay is the reverse current relay. It connects and disconnects the battery from the generator. If the generator is not disconnected from the battery when the engine is not running, the battery will discharge itself through the generator which will try to act like a motor. There are two coils, inner and outer. The first coil, which is covered by the second winding, is formed by many turns of very fine magnet wire. One end is attached to the armature terminal and the other is attached to the base of the regulator. This is a shunt winding. The second winding is made from heavy magnet wire and has only a few turns. One end is attached to the wire running to the current limiting circuit from armature terminal and the other end is attached to the movable contact. When the points are closed, the circuit is completed at the *Bat* terminal, connecting the battery to the generator.

The current limiter has but one coil composed of the heavy magnet wire which connects the *A* or *Gen* terminal to the reverse current relay. Depending upon the application, the wire may have as few as eight turns or as many as twenty.

The voltage control consists of a single coil consisting of many turns of very fine wire connected at one end to the *A* terminal and grounded to the base at the other.

The contacts of the reverse current relay are normally open. The magnetic field generated by the fine wire winding pulls the moveable contact

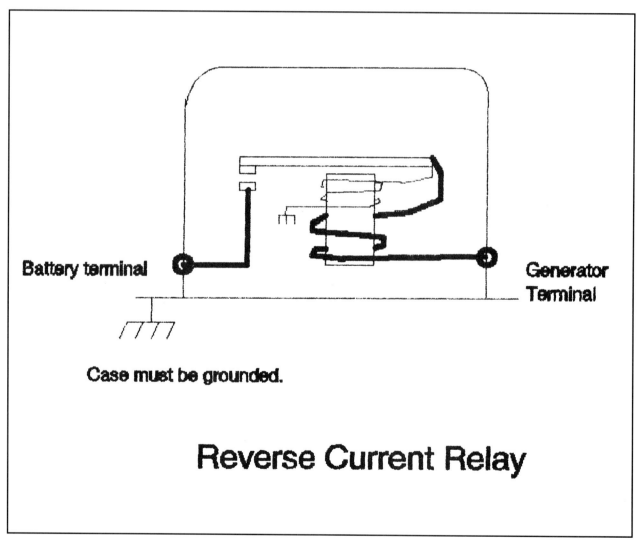

Battery terminal

Generator Terminal

Case must be grounded.

Reverse Current Relay

The reverse current relay is also known as a cutout. It disconnects the battery from the generator when the generator voltage is less than the battery voltage.

armature against the resistance of a spring until contact is made. The flow of current through the heavy winding as the generator provides power to the system intensifies the magnetic field and insures that the contacts are held together firmly.

The other two sets of contacts are normally closed. The field winding is connected directly to ground. When the voltage of the system increases to the set point of regulation, the voltage limiting contacts are pulled open. An adjustable spring provides resistance until the magnetic field is strong enough to pull the contacts open. The resistor provides a path to ground. This prevents excessive arcing at the contacts as well as reducing the flow of current through the field coils.

If the demand for current exceeds the design capability of the generator, the current flowing through current limiting relay coil creates a mag-

netic field of enough strength to overcome the resistance of the adjustable spring. The points open, reducing the flow of current through the field coil. Of course, as soon as the points open, the current through the field coils decrease. The magnetism within the generator decreases, reducing output. Voltage, or current decreases allowing the open contact to close. Current flow increases, magnetism increases, output increases, and the cycle repeats itself.

If the voltage is set too low, the tension upon the adjusting spring must be increased. Greater tension means that the voltage must rise higher before the magnetic field becomes strong enough pull the regulator armature down, opening the points. Likewise, if the voltage is too high, tension on the adjustment spring can be reduced, allowing lower voltage to open the points. The same is

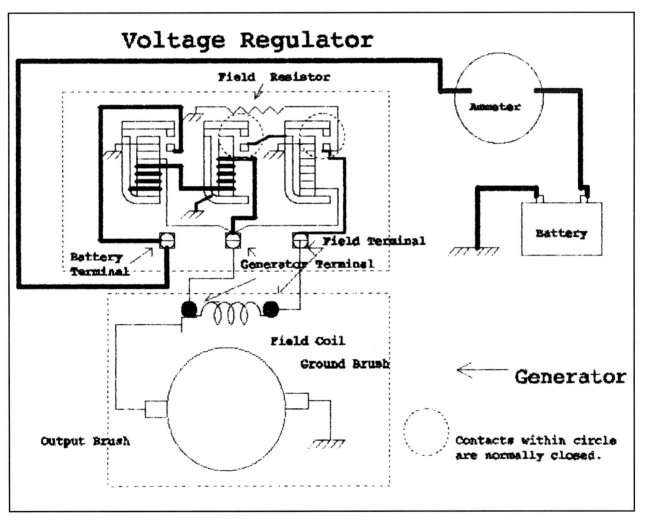

The three-unit voltage regulator does three things. It disconnects the battery from the generator when the generator voltage is less than battery voltage (reverse current relay). It limits the amount of current that the generator is allowed to produce (current limiting relay). It controls voltage so that the battery is properly charged (voltage control relay).

true of the current limiter; higher tension means greater current flow. Reverse current relays rarely require adjustment. All they have to do is to connect the generator to the battery, when battery voltage is reached. When the generator voltage is lower than the battery voltage, they must disconnect the battery from the generator. When the current through the winding reverses as the engine is shut down, the magnetic field repels the relay armature, forcing it to open.

The Testing

Now to the nitty-gritty. If the battery is good, and fully charged, the voltage should be regulated. With your finger, increase tension on the voltage control relay armature (press in the same direction as the adjustment spring is pulling). Voltage and amperage should both increase. If they don't increase, try speeding up the engine. Try again. If they still don't increase, try adding pressure to the current limiter at the same time. If you still can't get additional output, examine the generator to find out why.

If the voltage is slightly higher than you wish by one or two tenths, shut off the engine, install the voltage regulator gasket, and cover and recheck. If the voltage is about where you want it, leave well enough alone. Installation of the cover will frequently change the setting by one or two tenths of a volt. Usually the setting decreases when the cover is installed. Stop the engine, remove the cover, and restart.

If you have a carbon pile, attach it across the battery terminals. Slowly decrease resistance (increase the load) until the voltage begins to fall. The current indicated by your test ammeter is the limiter adjustment. If it is too high, decrease the tension on the current limiter adjustment spring and go through the cycle again. Do not leave the carbon pile turned on for long periods. If the current limiter is set too low, increase the tension until you reach the desired reading.

If it seems like you are having to increase the tension more than seems normal, stop adjusting. Momentarily, with your fingers, increase the tension upon both voltage and current limiter springs. If both voltage and amperage do not increase, speed up the engine and try again, as you did when setting the voltage control. Once you are convinced that you have reserve power, return to the adjustment task.

If you have a slipping fan belt, for instance, no amount of adjustment will increase voltage or amperage. The same is true if you are attempting to get forty amps out of a twenty amp generator.

When you make your adjustments, remember to adjust voltage so the battery remains fully, but not over, charged. Adjust the current limiter to

A typical tractor voltage regulator. When used with a third brush generator, this unit will sometimes incorporate a fuse from inside the regulator to ground. If the connection from regulator to battery or from battery to ground is lost, the fuse will blow, preventing damage to the generator.

the specified maximum output of the generator. If you exceed the capability of the generator to dissipate the heat produced in the armature, the temperature will rise until the generator fails (it will literally go up in smoke).

Frequently, you will discover that someone has already worked on the regulator. Most of the time they have screwed everything up so that you have to start from scratch. This is easy enough to do if you have all the specs. If you don't have the specifications, you must use the old system of trial and error.

Back to Earth

Returning to the Earthworm tractor, remove the cover and gasket. Very fine sandpaper and a drop of solvent will usually clean the points. Check your work with the multimeter. Experience has taught me that the center or current limiter relay points cause most of the trouble in three-unit regulators. With the multimeter connected, try tapping your finger on the armatures. If they return to zero resistance consistently, it is worthwhile to install the regulator without the cover for a test run. Make the test run long enough to get the regulator hot. Look for steady output. Look at the points during operations. There should be very little arcing at the points. Be sure to apply enough load to the battery to get the current limiter into operation. Run several cycles.

Before I bought an oscilloscope, I used my old aircraft radio headphones to listen to the action of the points. A good voltage regulator sounded smooth and regular, a bad regulator would give me a headache. On an oscilloscope, poor contact is revealed by lots of "hash" or "noise" instead of a nice square wave.

I used to maintain alternators, starters, and voltage regulators for a fleet of eighty or so large Ford cement trucks. When their mechanics had a charging system problem, they would replace the voltage regulator first thing. If that didn't cure the problem they would replace the alternator. The old voltage regulators were thrown into a bushel basket. When the basket was full they would bring it to me so that I could test and repair the whole works. I set up my test bench with a Ford alternator and my 'scope. Ford or Motorcraft always riveted the covers in place so I would run the test before I attempted to remove the cover. The regulators with proper voltage setting and a clean pattern would go directly into the "out" box. Regulators with a bad pattern went directly into the scrap barrel. I opened up regulators with no output. In most cases the field wire had been grounded with the switch on. This caused a fuse wire inside to fail. After a quick solder job they were ready for the test.

One day the fleet's head mechanic was watching me test the regulators and I was explaining how I passed or failed them depending upon the appearance of the regulating wave form. "What you really like to see is the next

The typical tractor voltage regulator with the cover removed.

A three-unit voltage regulator. The reverse current relay is always the relay with the largest set of contacts. The current relay is in the center. The voltage control relay is at the left.

thing to a perfect square wave," I was saying as I plugged another regulator.

"Oh," Al, the mechanic said, "that looks like a pretty square wave to me." He pointed to the screen. There was a perfect example of square wave if I ever had one.

"That one doesn't look natural," I said, "lets open it up." The cover in this case was held on with a pair of sheet metal screws. I opened it up. The familiar voltage control relay was missing! In its place was small circuit board with a small potentiometer for easy adjustment of voltage. It still had a mechanical relay to turn off the idiot light.

Chapter 18

Electric Starter Motors

An electric starter motor consists of six major components: A steel field frame with field coils and pole shoes; an armature which is usually wound with square wire making a single turn from one commutator bar to the next; a com end frame; drive end frame; and some form of drive mechanism to connect the starter to the engine starter ring gear.

Electric starter motors may use two, four, or six poles. Two-pole starter motors were used on some early automobiles and by some later-model small air-cooled engines used on lawnmowers and garden tractors. The four-pole starter is the most common. Larger diesel engines may use six-pole starter motors.

Starter motors are usually series motors. That is, the electrical current flows from the input terminal through the field coils and then through the armature to ground. Some motors have the current flow from the input terminal to the armature and from the armature to the field coils and from the field coils to ground. As long as the two components are used in series, it makes no difference in operation. Series motors have good starting torque. They accelerate until the load and output of the motor is in balance.

Shunt Coils

Many starter motors have one or more field coils as shunts. That is, the shunt coils have one end of the winding attached to the input terminal and the other end goes to ground. The shunt coils are made from many turns of fine wire. Current flowing through the shunt winding creates a stronger magnetic field. The armature windings passing through the stronger magnetic field cause a reverse electromotive force to be generated in the armature. The voltage generated reduces the flow of current through the armature. This reduces the strength of the magnetic field doing the work. Power is reduced as is the consumption of electricity. The use of shunt coils enable the man-

ufacturer to reduce the number of starter parts it has to make to cover the range of engine sizes. The housing, armature, and drive assembly remain the same for engines ranging in size from 400ci down to 200ci. A large engine would use two pairs of series-wound field coils. A medium engine might use two series-wound coils (two pole shoes would have no field coils). A small engine would use one pair of series-wound coils and two shunt coils. The large engine would use a large

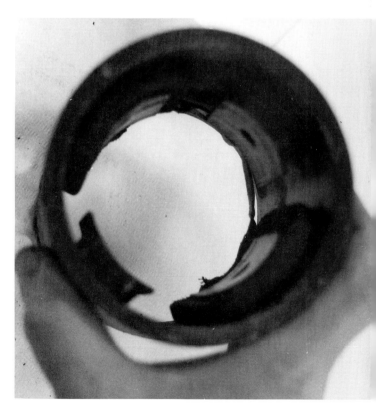

A field frame with two series coils and no shunt coils. Two of the pole shoes have no coils at all. The four-pole shoes make this a four-pole starter.

battery and the small engine would use a much smaller battery. The net result was variety with a minimum number of parts.

At the business end of the magneto is the drive assembly. An early manufacturer of starter drives was the Bendix Corporation. Bendix drives were inertial; the gear rides on a spiral shaft, inertia prevents the drive from turning until it reached the end of the spiral and engaged the flywheel gear. They featured a flat coiled spring to absorb starting shock. In the same way that electric refrigerators became fridges, after the General

A Bendix drive. The heavy flat-coiled spring behind the drive gear is the Bendix spring. The spring takes up the shock when the drive gear engages the ring gear on the flywheel.

A Folo-Thru Bendix drive. The design of this drive solved the problem of premature disengagement of the gear from the flywheel suffered by early Bendix drives. The drive is locked in the extended position until it reaches high rpm. If the engine only fires a couple of time, the drive won't disengage.

Electric Frigidaire, all starter drives became known as bendixes. General Motors used an over-running clutch assembly which was engaged by a lever operated by either the driver or by an electric solenoid. Both systems are found on old tractors.

My 1936 Hudson Terraplane had a Bendix drive. On cold days in Minnesota it revealed the two faults of the design. Cold grease on the spiral shaft prevented the gear from sliding forward and engaging the flywheel gear. There was an access plate under the hood so that I could get out and turn it by hand until it engaged. I would climb back into the car and hit the switch. The engine would turn until it fired one cylinder which caused the drive gear to disengage. I would get out and re-engage the gear by hand and try again. How I welcomed the coming of spring. Had I any sense at all, I would have washed the business end of the starter with gasoline or solvent.

In the 1950s, Bendix came up with a follow-through which remained engaged until the engine was running and spinning the drive fast enough for a pair of locking pins to release and allow it to retract. For sheer durability, the old Bendix drive had no peer. For convenience, the over-running clutch type took first place.

An overrunning clutch drive. This system uses either a solenoid or a mechanical lever system which engages the drive before energizing the starter.

Disassembly

The first thing I do when I take a defective starter apart is to give the innards a good sniff. The acrid smell of burned varnish means that major damage has been done. The pleasant perfume of old oil and grease usually means that the starter is repairable. Remove the two through bolts and the com end plate. Examine the commutator and brushes for uneven wear patterns. If there is tearing or uneven wear of commutator bars, examine the armature for melted and thrown solder. Thrown solder appears as a line of gray dandruff opposite the rise of the commutator. If dark voids appear where the conductors are soldered to the bars, the armature should be tested, cleaned, and soldered.

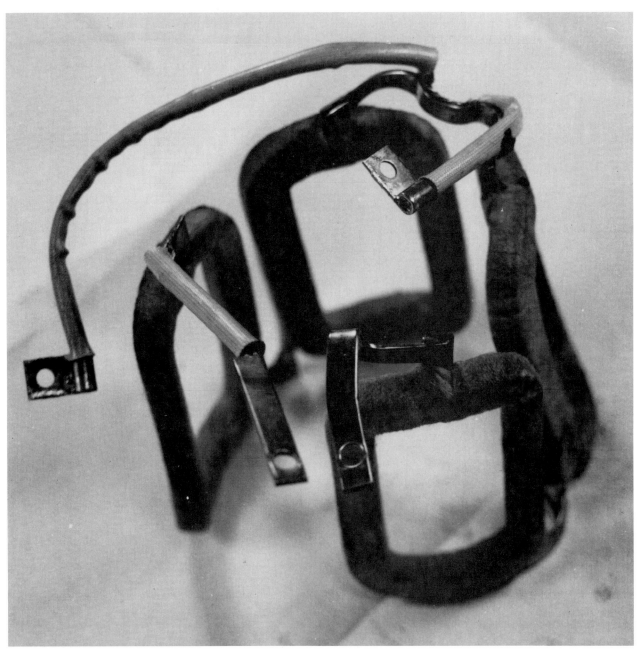

Two pair of series-wound coils. The two flat tabs at the bottom of the picture conduct electricity from the solenoid to the coils. After going through the coils, electricity goes through the two wires at the top of the picture to the insulated brushes. This makes a powerful starter.

Remove the drive end housing, drive, and—if there is one—the center plate. Note the location and size of the washers; they must be installed properly when the starter is reassembled. The drive may be held in place with a woodruff key and a set screw or a solid or roll pin through the shaft. Examine the teeth of the drive. If teeth are broken or missing, the drive will have to be discarded. If the teeth are rounded or blunted, the proper shape can be restored by careful grinding.

Examine the drive end housing for bushing wear. If the shaft is worn, undersized bushings are available. A friendly rebuilder may be helpful.

Field Coils

Pay particular attention to the attachment of the field coils to the terminal stud. Use a screwdriver to pry on the coil at a soldered joint. Frequently, the solder will have fractured at this point, leaving only a single pair of field coils in operation. With the brushes free in air and not touching anything, test the field coils for grounding. If there is continuity where there should be none, the field coils will have to be removed and checked. Replace the wrapping if the material has deteriorated badly. Rebuilder supply companies, such as Ace Electric in Kansas City, IPM in Chicago, and Rebuilders Supply in Miami, stock fiberglass, cotton, or paper tapes for this purpose. You may have to order through a rebuilder. Companies that supply materials for electric motor repair also stock such materials.

Armature and Commutator

I clean the armature with stoddard solvent, rinse it in hot water and dry it with compressed air. I then glass bead blast the commutator and the conductors where they connect to the commutator. A large electric soldering iron, non-corrosive flux, and rosin-core solder do the job. If you

Field frame. Note how the ends of the coils are soldered in the slot in the terminal. Always check these for broken connections. I pry with a small screwdriver, and look for any movement. It takes a large soldering iron to make a good solder joint.

have a solder pot, flux the parts to be soldered, grease the parts not to be soldered, and submerge the end of the armature in solder, being very careful not to solder beyond the end of the commutator.

Turn, undercut, and polish the commutator and completely test the armature with the growler for shorts, opens, and grounds. Install the center plate and the drive assembly. Place the armature in the drive end housing and secure the center plate, if there is one. Install the refurbished field frame, with new brushes, if required, and com end plate. Insert the through bolts and tighten. I always chuck the starter in the vise and apply power to the terminal. Starters with inertial drives do tend to vibrate. I listen for rattling and check the drive end bushing for excessive side play (there should be none or very little). I like to force a piece of oak against the drive gear when I am testing the starter. If I am able to stop it from turning, it's time to go back to the old drawing board. On the other hand, if the room fills with wood smoke, I calculate that the starter will operate in a satisfactory manner.

Occasionally, when you install a bushing in a housing, it turns out to be a little snug. When you

Commutator end plate with brushes. The brushes conduct electricity from the field coils to the armature commutator. Two brushes are grounded. After the current flows through the armature, it goes through two brushes to ground. Always check the brush holders for free movement. Look at the face of the brushes where they rub on the commutator bars. They should be highly polished. If a brush face is dull or dark when compared with its neighbor, it has poor contact. Overhaul specifications will usually give brush spring tension specifications. It is measured with a spring scale at right angles to the brush holder.

apply power to the starter, at first it sounds all right. Then the starter slows down and smoke begins to pour from the housing. You have two choices. Apply more power until the bushing burns in, swap the bushing for a larger one, or try to ream it a little. If you take the first choice and burn it in, be sure to take it apart when it cools and add a good lithium grease.

When checking the manual electrical switch, remember that bright-colored pitting is what you want to see. If the points are worn or are dark and slimy, replace them.

The Solenoid

Delco solenoids can be opened up for examination by removing the nut on the terminal that connects the solenoid to the starter terminal. Take off the nuts on the small terminal stud marked with an *S*. Unscrew the two screws that hold the cap to the housing. Remove the cap while pushing the two terminal studs back through the cover. The input terminal stud is the one that wears. Loosen the nut and push the stud back into the

The business end of the starter switch. A bar with angled ends is forced against both the terminal at the starter and the terminal connected to the battery. The bar conducts current from battery to starter. The ends of the bar should have bright pits where they connect to the terminals. The contact areas of the terminals should likewise have bright pits. A rough appearance is okay. Dark, slimy points or contacts are bad!

Armature with key slot and a hole for a set screw. Commutator wear pattern indicates very poor brush contact.

An International Harvester drive end housing. The bushing at the end of the starter housing should be replaced whenever you can wiggle the armature in the hole. A worn bushing will cause a rattle or knocking sound when the starter is operated under no load in a test stand or workbench vise.

cap. Turn the stud one hundred-eighty degrees. Clean the disk. Again, bright pitting is a sign that the disk is making good contact. If there is a dark, worn spot on the disk, reverse the disk if possible. Otherwise replace the disk contact or the complete solenoid.

If the starter doesn't want to spin briskly when you apply power, but you get a very heavy draw, try removing the through bolts, one at a time, while testing. If the starter operates with one through bolt removed, take the com end plate off and check for possible short circuits.

Examine the through bolt carefully; sometimes a small burn spot will indicate the location of the short. Short circuits often occur when the connecting straps between the field coils touch ground.

Be sure to clean all nuts and terminal studs with a wire brush. It is also a good idea to clean all battery connections and ground cable connections.

Chapter 19

Tractor Wiring

One of my jobs, after my company started manufacturing wire harnesses, was to set up the machines that installed terminals on wires. The machine was a two-ton punch press with special dies for crimping the terminals. The terminals came on large spools and were fed automatically. You poked the stripped wire into place and touched the foot control and the machine crunched it into place. Each different terminal and each wire size required a final adjustment to the machine. I would install a terminal on a wire and clamp the wire into a machine which measured the force required to cause the wire to break or the terminal to come loose. If the terminal came off before the specified pull, I would tighten the adjustment on the die set. If the wire broke before the specified force, I would loosen the adjustment.

The ideal setting was one that allowed the terminal to pull free just before the wire broke. The force required was substantial. I have long since forgotten the specifications, but it seems to me that a fourteen-gauge wire with a ring terminal took around a hundred pounds. The sales representative maintained that the terminal was pressure-welded to the wire. I remember cutting a terminal into two parts at the joint and being unable to tell where the wire ended and the terminal began.

It is not economically feasible for the antique tractor repairer to buy a two-ton punch press and fifteen or twenty die sets, but a good set of terminal pliers and an assortment of terminals is a worthwhile investment.

Terminals

Terminals are either pre-insulated and uninsulated. The nice thing about pre-insulated terminals is that they are color coded. A red sleeve indicates the terminal is for sixteen- or eighteen-gauge wire. Blue is for sixteen- and fourteen-gauge. Yellow indicates twelve- and ten-gauge. Uninsulated terminals may or may not have the wire and ring size stamped on the tongue.

Terminal installation tools can be simple or compound. Simple terminators are like a pliers; you grab the terminal and squeeze. Compound terminators use an extra set of joints and (usually) longer handles to exert very great pressure. The best type use a ratchet affair which will not release until they are completely closed. This prevents terminals from being only partly compressed. Compound tools tend to be a little on the pricey side. The inexpensive Amp Champ will do a satisfactory job if you squeeze it tightly.

The weakest part of a wire with terminals is at the joint of the terminal and wire. Pre-insulated terminals require two strokes. The first stroke crimps the metal barrel of the terminal to the wire. The second stroke crimps the insulation tightly around the insulation of the wire. The part of the tool for insulation crimping is usually marked *Ins* and the opening is round.

Remember to lightly scrape the wire before installing the terminal. A dirty connection is a bad connection.

Choosing Replacement Wire

If you are restoring your old tractor to a condition of pristine beauty, the wire of choice is rubber-insulated with a varnished cloth outer cover. The terminal is a solder type with a rubber sleeve over the joint. Companies which sell materials for automotive restoration will usually stock this type. Again, cleanliness is next to godliness. Scrape the wire with a dull knife before you solder it in place. A large soldering gun is well suited for this type joint. Use rosin core solder.

When my old company started to make wire harnesses for Minneapolis-Moline, they specified rubber sleeves and a tin-dipped joint. We would install the terminals with a pair of common pliers and dip the terminal first into liquid solder flux and then into the solder pot. Not the fastest

The best terminating tools have a compound operating lever.

system, but I'll bet the terminals never fell off the wires.

For repairing old tractors, I recommend stocking three sizes of wire. Fourteen-, twelve-, and ten-gauge will take care of ninety percent of the repairs. Fourteen-gauge wire works for a load of ten amps (six volt system) or twenty amps (twelve volt system). Use twelve-gauge wire for loads of ten to twenty-five amps (six volt system), or twenty up to fifty amps (twelve volt system). Twenty-five to fifty amps (six volt system) or fifty to seventy five amps (twelve volt system) calls for a ten-gauge wire. I have based these figures on a ten-foot-long wire in a grounded system. When in doubt, use heavier wire.

Most electrical devices have the power requirement listed. If the listing is in watts, divide watts by system voltage. For example, a one hun-dred-watt quartz-halogen lamp would draw eight and one-third amps when used on a twelve-volt system. A fourteen-gauge wire would handle a pair easily. A pair of one hundred-watt quartz lights on a six-volt system would require twelve-gauge wire. (16.6 x2= 33.34 amps.)

Converting to a Twelve-Volt System

Converting an old tractor from six to twelve volts requires no change in wiring since the six-volt wiring is quite heavy. Converting a six-volt generator to twelve volts requires a change of field coils in the generator. The voltage regulator must be replaced by a twelve-volt regulator. The six-volt light bulbs must be exchanged for twelve-volt bulbs. Switches should be okay. Install a twelve-volt battery and you are ready to go. The ammeter will work fine. The starter and solenoid will be all right. The starter will really crank the engine fast. If you have (heaven forbid), a battery ignition system, install a standard ballast resister between the ignition switch and the coil. If you have an electric solenoid on the starter, replace it with a twelve-volt solenoid and install a wire from the *R* terminal to the input side of the coil. If the starter uses a mechanical switch, replace the switch with one that has an extra terminal on the side to provide direct power to the coil when cranking the engine.

Installing Lights

If you decide to festoon your tractor with lights and find that the load is too great for the old generator, consider installing an automotive alternator in its place. The easiest change is to a Delco model 10SI. Ask your friendly automotive rebuilder, or parts jobber, for a "one wire" set up. The regulator is part of the alternator, but doesn't require any external hook up. Just run a ten-gauge wire from the output terminal on the alternator to the ammeter. If you haven't got an ammeter, run directly to the battery. Gun the engine once after you start it, and the alternator should produce plenty. JC Whitney advertises one wire Delcos in their catalog for fifty to sixty dollars.

Finally, this recommendation: When you install a wire, be sure that there is no tension on the terminal. The best way is to tape or tie wrap the wire to the structure about three to six inches from the terminal. Leave a little loop in the wire to be sure that there is no tension on the terminal.

If you decide that you should use a ten-gauge wire for your new one-wire alternator system, but you have only twelve-gauge wire on hand, you can use two twelve-gauge wires in parallel. I prefer to use the proper wire, but in a pinch I have used parallel wires with no problem.

Index